SEMICONDUCTOR SUPERLATTICES

Growth and Electronic Properties

SEMICONDUCTOR SUPERLATTICES

Growth and Electronic Properties

Editor

H T Grahn

Paul-Drude-Institut für Festkörperelektronik
Hausvogteiplatz 5-7, D-10117 Berlin
Germany

World Scientific
Singapore • New Jersey • London • Hong Kong

Published by

World Scientific Publishing Co. Pte. Ltd.

P O Box 128, Farrer Road, Singapore 9128

USA office: Suite 1B, 1060 Main Street, River Edge, NJ 07661

UK office: 57 Shelton Street, Covent Garden, London WC2H 9HE

ISBN 981-02-2061-8

Printed in Singapore by Uto-Print

To

my mother, my father,

my sister, and my brother

PREFACE

The subject of semiconductor superlattices has reached a rather mature state 24 years after its first proposal by L. Esaki and R. Tsu. This book focuses on the electronic properties of semiconductor superlattices. In particular, the transport properties of this system under the application of an electric field parallel to the superlattice direction are discussed. Two very old theoretical predictions have been recently realized in semiconductor superlattices, the Wannier-Stark localization and the Bloch oscillations. Furthermore, semiconductor superlattices can exhibit several regions of negative differential resistance with a different physical origin than in bulk semiconductors, i.e., miniband transport and resonant tunneling versus Γ-L transfer. The interaction of this non-linear transport characteristic with a large carrier density leads to a very peculiar current-voltage characteristic, which is shown on the book cover.

This book is divided into five chapters. The first chapter discusses the fabrication and characterization of semiconductor superlattices. In the second chapter, the transport properties of superlattice minibands are reviewed. The localization of the miniband states with increasing electric field, i.e., the Wannier-Stark localization, and the very recent experimental observation of Bloch oscillations are the subject of the third chapter. At large electric fields, the states become completely localized within the individual wells of the superlattice. However, due to the coupling between adjacent wells, resonant tunneling between different subbands can occur in this field regime, which will be presented in chapter 4. Finally, the non-linear velocity-field characteristic of superlattices can result in very peculiar effects for large carrier densities. The last chapter contains an overview on electric field domains in semiconductor superlattices.

The editor would like to thank all the authors for their contribution. The scientific collaboration with many colleagues is greatly acknowledged. In particular, I would like to thank Fernando Agulló-Rueda, Gottfried Döhler, Dietrich Bertram, Luis Bonilla, A. Fischer, Jorge Galan, Jörg Kastrup, Robert Klann, Klaus von Klitzing, Si Ho Kwok, Boris Laikhtman, Roberto Merlin, Wolfgang Müller, Klaus Ploog, Frank Prengel, Wolfgang Rühle, Klaus Schmidt, H. Schneider, Eckehard Schöll, and Andreas Wacker for their many contributions and support. I am very thankful to Ilka Schuster, who photographed all the illustrations, and Doris Spaniol, who put the illustrations into the right place.

Finally, I would like to express my gratitude to my companion Astrid Gollhardt for supporting me and being patient with me during the final stages of the completion of this book.

Berlin, November 1994 Holger Grahn

CONTRIBUTORS

Fernando Agulló-Rueda
Instituto de Ciencia de Materiales de Madrid (CSIC)
Campus de Cantoblanco C-IV, E-28049 Madrid, Spain

Jochen Feldmann
AG Festkörperelektronik, Fachbereich Physik
Philipps-Inversität Marburg, Renthof 5, D-35032 Marburg, Germany

Kenzo Fujiwara
Department of Electrical Engineering, Kyushu Institute of Technology
Tobata-ku, Kitakyushu 804, Japan

Holger Grahn
Paul-Drude-Institut für Festkörperelektronik
Hausvogteiplatz 5-7, D-10117 Berlin, Germany

Alain Sibille
Departement Electronique,
École Nationale Supérieure de Techniques Avancées
32, boulevard Victor, F-75015 Paris, France

CONTENTS

Preface vii
Contributors ix

CHAPTER 1. GROWTH AND CHARACTERIZATION
by KENZO FUJIWARA

1.1. Introduction 1

1.2. Classification of superlattices 2
    1.2.1. Superlattices types 2
    1.2.2. Tailoring the periodicity 8

1.3. Superlattice materials 10

1.4. Growth methodology 12
    1.4.1. Growth technology 12
    1.4.2. Thickness control 14
    1.4.3. Monolayer thickness control 15

1.5. Characterization 18
    1.5.1. Miniband structure 18
    1.5.2. Excitonic effects 22

References 26

CHAPTER 2. MINIBAND TRANSPORT
by *ALAIN SIBILLE*

2.1. Introduction: Miniband transport, dream or reality? . . . 29

2.2. Miniband structure: Calculation and experimental
observation 30
   2.2.1. Generalities on superlattice minibands 30
   2.2.2. Experimental determination of miniband parameters . . . 34

2.3. Semiclassical miniband transport models 38
   2.3.1. Esaki-Tsu model 38
   2.3.2. Relaxation time approximation 40
   2.3.3. Full Boltzmann equation 43

2.4. Experimental techniques providing evidence of
miniband transport 49
   2.4.1. Optical techniques 49
   2.4.2. Miniband resonances in transport 51
   2.4.3. Miniband transport in transistor structures 52
   2.4.4. Experimental demonstration of miniband NDV through
       d.c. current-voltage measurements in unipolar structures . 55
       2.4.4.1. Undoped superlattices 55
       2.4.4.2. Doped superlattices 59
   2.4.5. Magnetotransport measurements 60
       2.4.5.1. Magnetotransport in crossed fields 61
       2.4.5.2. Magnetotransport in parallel fields 64
   2.4.6. A.C. electrical characterization of miniband transport . . . 65
       2.4.6.1. Origin of the a.c. response 65
       2.4.6.2. Microwave probing 66
       2.4.6.3. Time-of-flight probing 69

2.5. Advanced analysis of miniband conduction 70
   2.5.1. Transport in wide minibands 70
   2.5.2. Transport in narrow minibands 73
   2.5.3. Miniband transport in the presence of disorder 78

2.6. Miniband conduction vs Wannier-Stark localization
and Bloch oscillations 82
   2.6.1. Miniband conduction vs Wannier-Stark localization 83
   2.6.2. Miniband conduction vs Bloch oscillations 87

2.7. Possible applications of miniband transport 89

   2.7.1. Millimeter/sub-millimeter oscillator sources 89

   2.7.2. Microwave/opto-electronic applications 91

2.8. Conclusions 93

2.9. Acknowledgments 93

References 94

CHAPTER 3. WANNIER-STARK LOCALIZATION AND BLOCH OSCILLATIONS

by *FERNANDO AGULLÓ-RUEDA and JOCHEN FELDMANN*

3.1. Introduction 99

3.2. Wannier-Stark localization 101

   3.2.1. Historical overview 101

      3.2.1.1. Stark ladder in the bulk 101

      3.2.1.2. Stark ladder in superlattices 101

   3.2.2. Properties at zero electric field 102

   3.2.3. Theory 104

   3.2.4. Optical experiments 108

   3.2.5. Localization length and coherence 112

   3.2.6. Excitonic effects 118

   3.2.7. Franz-Keldysh oscillations 122

   3.2.8. Miscellaneous work 124

   3.2.9. Electro-optic applications 125

3.3. Bloch oscillations 130

   3.3.1. Introduction 131

   3.3.2. Four-Wave mixing experiments 134

   3.3.3. Bloch oscillations in narrow minibands 136

   3.3.4. LO-phonon scattering in wide minibands 139

   3.3.5. Field-induced exciton ionization in a wide

      superlattice miniband 145

   3.3.6. Coulomb effects 147

3.4. Acknowledgments 148

References 149

CHAPTER 4. RESONANT TUNNELING
by HOLGER GRAHN

4.1. Introduction 155

4.2. Theoretical background 156

4.3. Experimental evidence for resonant electron tunneling 159
    4.3.1. Photocurrent-voltage characteristics 161
    4.3.2. Time-of-flight experiments 163
    4.3.3. Non-thermal occupation of higher subbands
        by resonant tunneling 167
        4.3.3.1. Infrared emission experiments 167
        4.3.3.2. Photoluminescence spectroscopy 169
        4.3.3.3. Electroluminescence spectroscopy 172
    4.3.4. Quantum cascade laser 178
    4.3.5. The quantum-well Pockels effect 180

4.4. Magnetotunneling of electrons 182
    4.4.1. Landau-level tunneling 183
    4.4.2. Tunneling in crossed electric and magnetic fields 189

4.5. Resonant tunneling of holes 194

4.6. Summary and conclusions 198

4.7. Acknowledgments 199

References 200

CHAPTER 5. ELECTRIC FIELD DOMAINS
by HOLGER GRAHN

5.1. Introduction 205

5.2. Theoretical model for domain formation 206

5.3. Current-voltage characteristics 212
    5.3.1. Doped superlattices 213
    5.3.2. Multistability 217
    5.3.3. Undoped superlattices 222

**5.4. Photoluminescence spectroscopy of
electric-field domains** 226
  5.4.1. Determination of domain field strengths 227
  5.4.2. Intensity dependence of domain formation 233
  5.4.3. Spatial distribution 235

5.5. Dynamics of domain formation 240
  5.5.1. Formation time 240
  5.5.2. Damped oscillations of the photocurrent 241
  5.5.3. Self-oscillations of the current 245

5.6. Γ-X tunneling and domain formation 247

5.7. Summary and conclusion 247

5.8. Acknowledgments 248

References 248

CHAPTER 1

GROWTH AND CHARACTERIZATION

by KENZO FUJIWARA

1.1. Introduction

Semiconductor superlattices can be defined as a new class of semi-conductor materials with a periodic arrangment of the constituents in such a way, which does not exist in nature. In this chapter, we will describe the semiconductor materials and their growth technologies which are currently used for the investigation of semiconductor superlattices. Since the first proposal by Esaki and Tsu[1] of synthetic artificial *superlattices* in 1970, great advances in the physics of such ultra-fine semiconductors, presently called *quantum structures*, have been made within the past two decades. The concept of quantum confinement, which has led to the observation of quantum size effects, was studied by Dingle et al.[2] in isolated quantum well heterostructures and is closely related to superlattices through the tunneling phenomena. Therefore, these two ideas are often discussed on the same physical basis, but each field has its own intrigue and different physics useful for applications in many electronic and optical devices.

To fabricate the specially tailored semiconductor structures which are being controlled on an atomic scale, sophisticated epitaxial growth technologies had to be developed. These advanced technologies allow us to prepare, for example, semiconductor ultra-thin heterostructure layers as small as one mono-molecular layer (the thickness is only 0.283 nm in GaAs along a cubic [001] direction). One of the representative epitaxial methods so far developed is molecular beam epitaxy (MBE), which was pioneered by Cho and Arthur.[3] *Epitaxy* stands for a growth method of thin crystal layers, which keeps the same lattice structure as that of a given substrate crystal (usually, the zinc-blende type crystal lattice structure). After the breakthrough of MBE technology, a variety of more refined growth technologies and new growth modes have been proposed and pursued for specific materials and applications.[4,5] This area of material science is still rapidly expanding and advancing, and strong efforts have been made to improve and optimize the ultimate growth control for further ultra-fine semiconductor materials such as quantum wires and quantum dots. Many interesting aspects of this active area of research are still developing further, but their basic ideas can be found in excellent review papers by Göbel and Ploog,[6], Parker,[7] and Ploog and Döhler.[8] For more general growth problems the reader should consult these articles.

The purpose of this chapter is to give a brief overview of semiconductor superlattices in general, how they are prepared and characterized. In section 1.2 we firstly classify the different types of semiconductor superlattice structures studied so far. The semiconductor materials used for the superlattice growth are described in section 1.3 followed by the growth methodology, which is discussed in section 1.4. Examples of actually synthesized semiconductor superlattices are surveyed in section 1.5 with special emphasis on GaAs/AlAs semiconductor superlattices, providing the present status of superlattice growth and discussing the remaining material problems for future studies.

1.2. Classification of superlattices

1.2.1. *Superlattices types*

Artificial man-made superlattices are different from bulk materials because new periodic potentials are incorporated in the lattice structure. The periodic potential can be introduced by doping or compositional modulation, i.e., there are two types of superlattice, doping superlattice and compositional superlattices. In Fig. 1.1 examples of the compositional modulation for an idealized atomic arrangement in a bulk semiconductor (a), a quantum well (b), and a superlattice (c) are schematically shown. For the zinc-blende type lattice structure the sublattice consists of two constituent elements (shown by large and small open dots), which provide the compositional modulation, while the other sublattice contains a common element (shown by small solid dots).

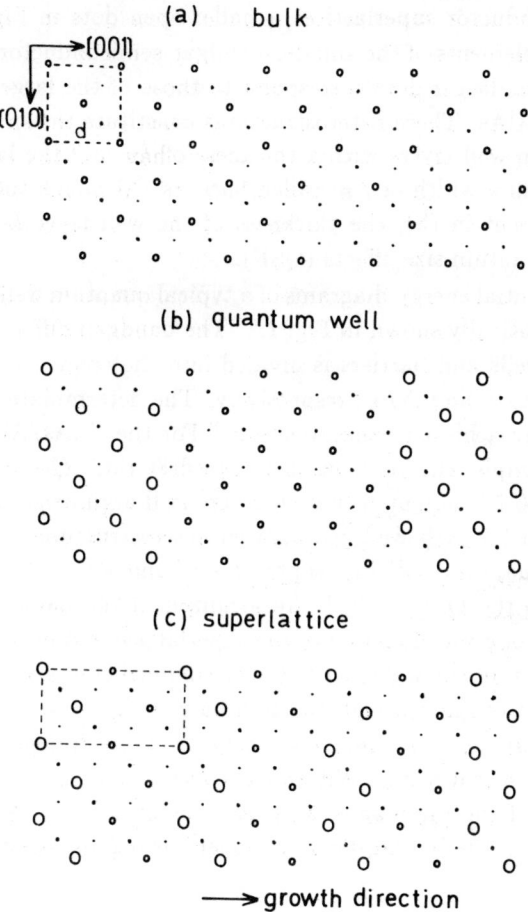

Fig. 1.1. Atomic structures of a bulk semiconductor (a), a quantum well (b), and a superlattice (c) with the zinc-blende type crystal lattice. The arrangement is modified along the [001] growth direction in (b) and (c). The superlattice example (c) has a periodicity of twice the original lattice constant d. Open and solid dots indicate elements of the sublattices.

A new periodicity is introduced in (c) for the sake of illustration with two molecular spacings along the [001] growth direction. As a result, the super-lattice unit cell is expanded to a size of twice the original unit cell along the z-axis. One molecular lattice spacing is equal to half the lattice constant, i.e., the size of the original unit cell, d. Since we are interested in III-V compound semiconductor superlattices, smaller open dots in Fig. 1.1 may be considered as elements of the smaller-bandgap semiconductor material, say GaAs, while the larger dots correspond to those of the larger-bandgap semiconductor, AlAs. These heterostructures constitute the smaller-bandgap (E_g^W) quantum well layers with a thickness of L_Z and the larger-bandgap (E_g^B) layers with a width of L_B, called barriers. Although the artificial periodicity is absent in (b), the thickness of the well layer L_Z is very thin resulting in quantum size effects (QSE).

The potential energy diagrams of a typical quantum well and superlattice are schematically shown in Fig. 1.2. The bandgap difference $E_g^B - E_g^W$ between the wells and barriers is divided into the conduction and valence band offsets, ΔE_C and ΔE_V, respectively. The determination of the band offset is usually achieved experimentally.[9] For the GaAs/Al$_x$Ga$_{1-x}$As systems, for example, the conduction band offset ratio $Q_C = \Delta E_C/(E_g^B - E_g^W)$ of $0.6 \sim 0.7$ is accepted, but there are still arguments about the precise value even for such well investigated heterostructures. In the case of the isolated quantum well (a), confined subband states E_1, E_2, etc. are formed due to the QSE,[2] which are localized in the particular well regions. The envelope wavefunction of the superlattice, however, is delocalized and spread over all the wells, because the confined states are degenerate in energy and can couple through the thin barriers by tunneling. Depending on how easy this itinerant tunneling motion is, a one-dimensional miniband is formed with a width 2Δ. Within the effective mass approximation, the energy width of the minibands is easily calculated for GaAs/AlAs superlattices based on the Kronig-Penney model[10] using an eigenvalue equation given by

$$cos(K\,D) = cos(\alpha\,L_z)\,cosh(\beta\,L_B)$$
$$+ \frac{1}{2}\left(\frac{m_W^*\,\beta}{m_B^*\,\alpha} - \frac{m_B^*\,\alpha}{m_W^*\,\beta}\right)\,sin(\alpha\,L_z)\,sinh(\beta\,L_B)\,,$$

$$(1.1)$$

where

$$\alpha^2 = \frac{2\,m_W^*\,E}{\hbar^2} \qquad (1.2)$$

and

$$\beta^2 = \frac{2\,m_B^*\,(V - E)}{\hbar^2}\,. \qquad (1.3)$$

V denotes the band offset, $\hbar$ Planck's constant divided by 2π, D the superlattice period, K the reciprocal wavevector, and m_W^* and m_B^* are the effective masses of the well and barrier materials, respectively. The results of the calculations are shown, e.g., for electrons and heavy-holes in GaAs/AlAs superlattices in Fig. 1.3. We note that the width can be tailored by appropriately selecting the thicknesses of the wells L_Z and barriers

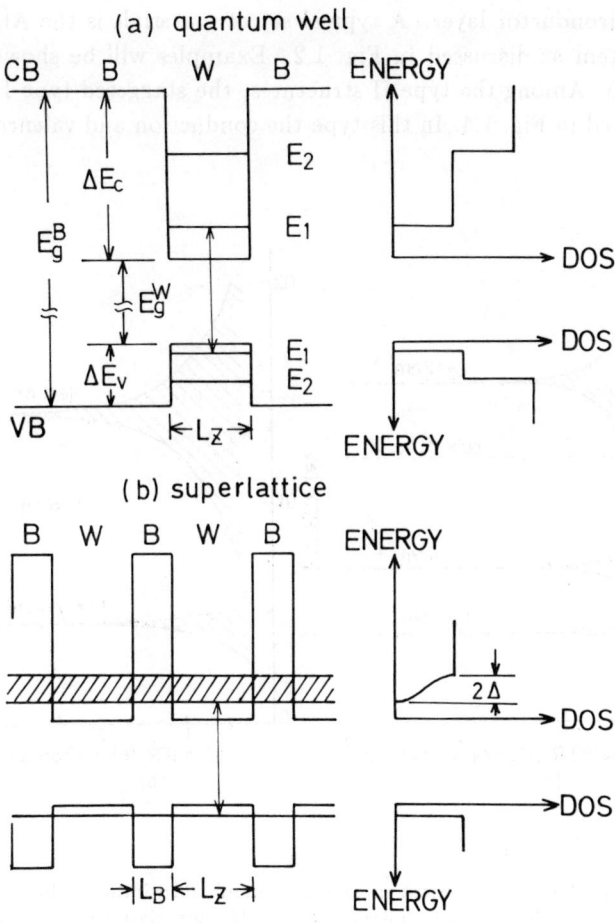

Fig. 1.2. Potential and subband energy diagrams of (a) quantum well and (b) superlattice.

L_B as well as the band offset. Comparing the electron and heavy-hole miniband, the width of the heavy-hole miniband is much narrower due to the larger effective mass. It is important to note that the miniband width can be easily tuned between zero and a few hundred meV.

According to the classification by Esaki,[11] the superlattice miniband structures are divided into two different types, called type I and type II as illustrated in Fig. 1.4. For the type I heterostructures the bottom of the conduction subband and the top of the valence subband are formed in the same semiconductor layer. A typical set of materials is the $Al_xGa_{1-x}As$-GaAs system as discussed in Fig. 1.2. Examples will be shown below in section 1.5. Among the type II structures, the staggered type II structure is illustrated in Fig. 1.4. In this type the conduction and valence subbands

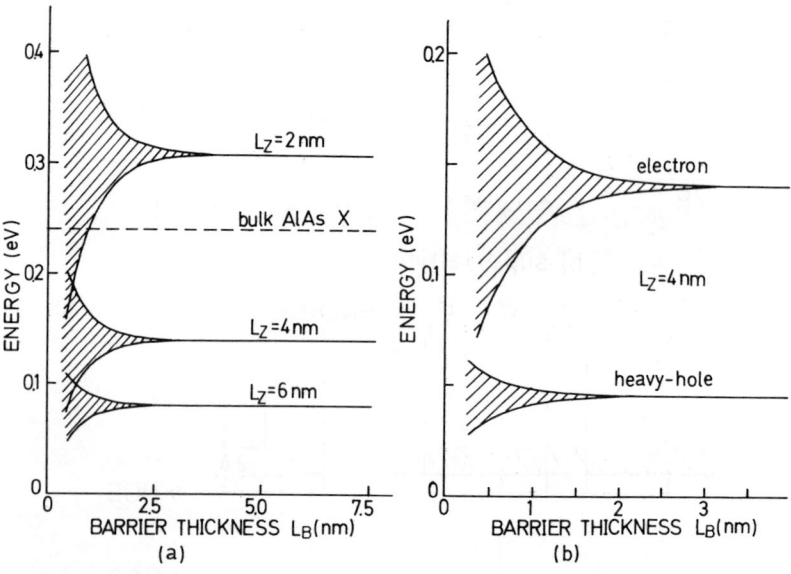

Fig. 1.3. Calculated confinement energies and miniband widths at the Γ-point for (a) electrons and (b) electrons and holes in GaAs/AlAs superlattices. The well width L_Z and the barrier thickness L_B are given in the figures. For the narrower well of $L_Z = 2$ nm the bottom of the electron miniband can be above the X-conduction band minimum of bulk AlAs when the L_B is increased (type II configuration).

are staggered in both real and reciprocal space, so that electrons and holes are confined in different layers. Optical evidence for the staggered type II structure was first given by Dawson et al.[12] in $Al_{0.37}Ga_{0.63}As$-AlAs quantum wells. For the $Al_xGa_{1-x}As/Al_yGa_{1-y}As$ ($x < y$) or AlAs system, the crossover between the direct Γ and indirect X minima can occur when the lowest conduction subband energy is raised by adding aluminium to the wells or by narrowing the well width (see Fig. 1.3). This is due to the fact that the X-minimum in the AlAs material exhibits the lowest energy of the X-conduction bands in $Al_xGa_{1-x}As$ ($0 < x \leq 1$). The all-binary GaAs/AlAs superlattices are interesting since both, the type I and the staggered type II structures, can be studied within the same set of materials just by changing the thickness of the well and barrier layers. For the monolayer $(GaAs)_n$-$(AlAs)_n$ superlattices the crossover from the type II to type I occurs at around $n = 12$ monolayers ($L_Z = L_B = 3.4$ nm).[13]

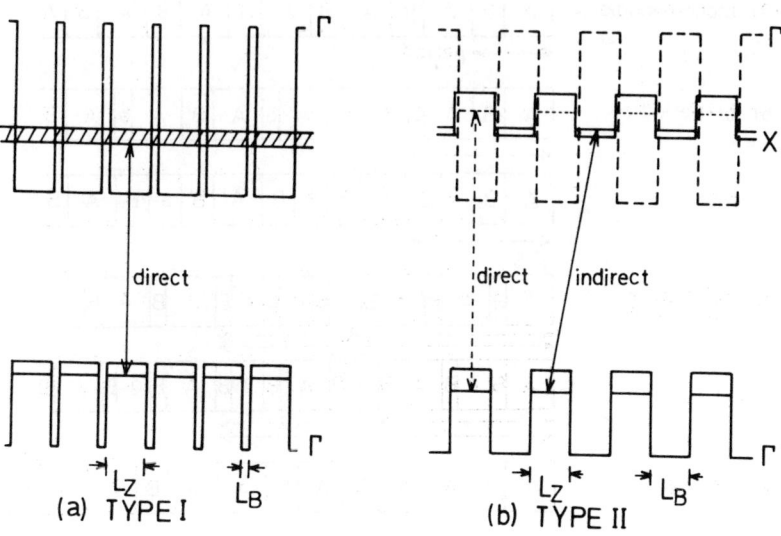

Fig. 1.4. Schematic energy diagrams of type I (a) and type II (b) superlattices.

1.2.2. *Tailoring the periodicity*

The artificial periodicity introduced in the superlattice structures can be tailored in various ways. In Fig. 1.1 we tacitly assumed mono-periodicity in the superlattice structure. However, there is some freedom in designing different types of periodicity. Due to the sophisticated material preparation technologies, new classes of periodic structures are in fact fabricated. Some examples,[14-16] which have their own unique structure, are schematically illustrated in Fig. 1.5. The simplest complexity added to the mono-periodic array is achieved in a superlattice with double-periodcity.[14] As examples of such structure, cross-sectional images taken by high resolution transmission electron microscopy (TEM) are shown for a double period GaAs/AlAs superlattice as well as for a mono-periodic superlattice in Fig. 1.6. In the superlattice in (a) the barrier thickness of AlAs (white) is alternating between 0.45 nm and 0.90 nm thickness with a constant GaAs (dark) well

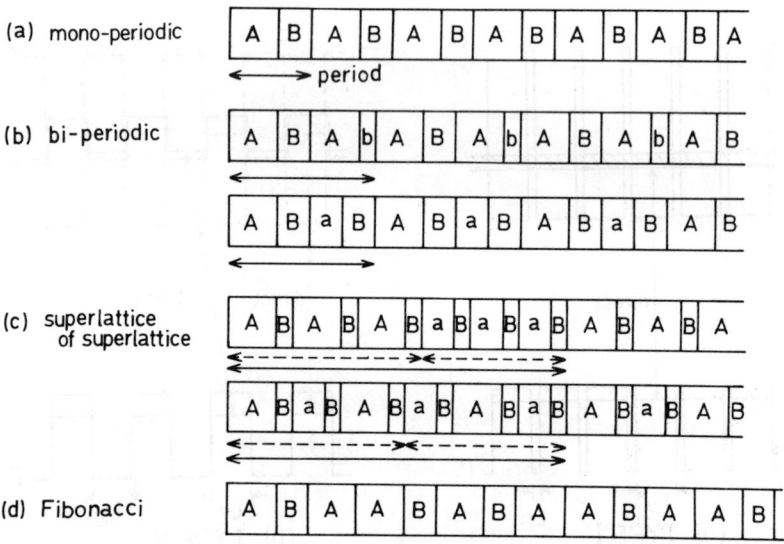

Fig. 1.5. Schematics of polymorphous superlattice structures: (a) mono-periodic; (b) two different types of bi-periodic structures; (c) two examples of superlattice of superlattice (SOS) structures; (d) Fibonacci sequence. One basic unit of A and B can be a well and a barrier or vice versa with compositional and thickness variations in (a) and (b).

width of $L_Z = 3.2$ nm. This intrinsic superlattice region is embedded in a p-i-n diode and exhibits interesting Wannier-Stark localization phenomena not previously observed in the usual mono-periodic superlattices.[14] These new phenomena are a duality of the optical transitions, and double-decoupling as well as resonant coupling between symmetric and anti- symmetric Stark ladders (see Chapter 3 for details).

Merlin et al.[15] have grown a one-dimensional quasi-periodic superlattice, in which alternating layers of GaAs and AlAs form a Fibonacci sequence, i.e., the ratio of the incommensurate periods is equal to the golden mean $\tau = (1+\sqrt{5})/2$. Using X-ray and Raman scattering experiments, the unique properties of this novel structure have been demonstrated. Recently, Einevoll and Sham[16] have proposed a new concept of superlattices of superlattices (SOS) for tailoring the unique subband structures suitable for infrared detectors using intersubband optical transitions. These new

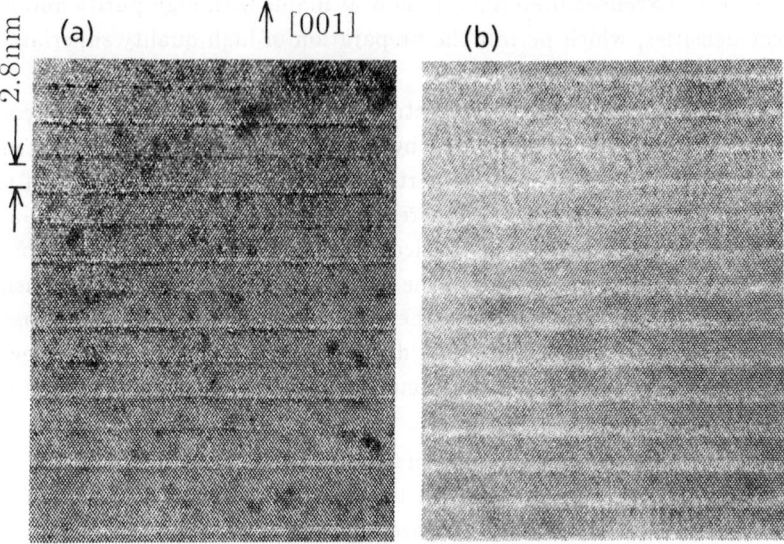

Fig. 1.6. TEM lattice images of GaAs/AlAs superlattices: (a) double period superlattice with two thicknesses (0.9 nm and 0.45 nm) of the AlAs barriers and a fixed width (3.2 nm) of the GaAs wells; (b) mono-periodic superlattice with fixed thicknesses of 0.9 nm for the AlAs barriers and of 3.2 nm for the GaAs wells.

types of heterostructures have been experimentally utilized, e.g., to fabricate the unipolar quantum cascade laser.[17]

1.3. Superlattice materials

Semiconductor materials, which are actually used to fabricate the superlattice structures, may be divided by the element groups, IV, III-V and II-VI. While the group III-V semiconductors have been extensively studied, group IV heterostructures such as the Si_xGe_{1-x} system are much more difficult to realize because of the large lattice mismatch. Nevertheless, the strain modification of the subband structures is interesting in these quantum structures and has attracted much attention.[18] The group II-VI semiconductor heterostructures have recently received a lot of attention because of the interests in blue light emitters.[19] Fig. 1.7 shows the relationship between the lattice constant and the lowest bandgap for typical semiconductors belonging to the IV and the III-V groups. These semiconductors, particularly the III-Vs, are important for applications in many optoelectronic devices. Excellent materials are now available with high purity and low defect densities, which permit the preparation of high quality superlattice samples.

As already pointed out in section 1.2, so far mostly the III-V compound semiconductors represented by the $GaAs/Al_xGa_{1-x}As$ heterostructures have been investigated. In particular, one distinguished merit of the GaAs/AlAs system is that the difference in lattice constant between GaAs and AlAs is very small (0.15% lattice mismatch) despite the highly corrosive nature of AlAs. In addition, the difference of their thermal expansion coefficients is also small. Thus, the strain remaining at room temperature can be minimized after cooling down from the higher epitaxial growth temperatures. The first compositional superlattice was realized using the $GaAs/Al_xGa_{1-x}As$ material system.[20]

Semiconductor superlattices may also be categorized by whether the lattice matching condition between the substrate and the epitaxial layers is satisfied or not. In order to prepare high quality superlattice samples in a reproducible manner, it is very important to have a good substrate material that can be used to grow lattice-matched, tailored hetero-epitaxial layers. This condition limits a practical choice of the semiconductor materials used for superlattices. Two kinds of substrates, GaAs and InP, are predominantly used (see dotted lines in Fig. 1.7). In most cases epitaxial

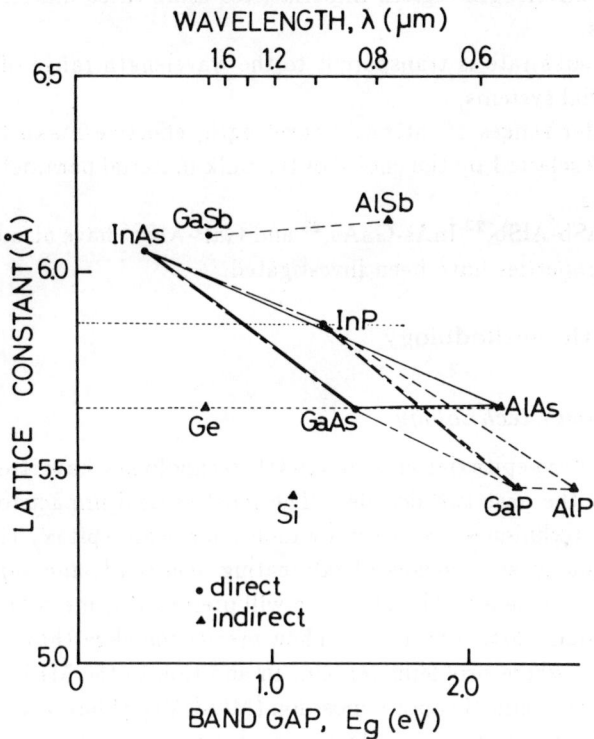

Fig. 1.7. Bandgap energy versus lattice constant for IV and III-V semiconductors at room temperature.

layers are prepared on homopolar [001] oriented substrates. Heteropolar substrates are more difficult for growth due to the complicated epitaxial growth mechanism. Besides $Al_xGa_{1-x}As$ epilayers, the growth of $In_x(Al_yGa_{1-y})_{1-x}P$ heteroepitaxial layers, which are useful for short wavelength laser applications, is possible on GaAs substrates. The ternary $In_xGa_{1-x}P$ and $In_xAl_{1-x}P$ epitaxial layers lattice-matched to GaAs show a spontaneous ordering in the group III sublattice.[21] These are natural superlattices, which depend on the growth conditions and substrate misorientation. $In_xGa_{1-x}As/InP$ as well as $In_xGa_{1-x}As/In_xAl_{1-x}As$ systems lattice-matched to the InP substrate are also well investigated. These quan-

tum heterostructures are important for lasers and modulators operative in
the 1.5 μm wavelength region. Superlattices using these materials have several merits:

(i) the InP substrate is transparent to the wavelength range of interest in
these material systems,

(ii) parameter ranges of interest (wavelength, effective mass, band offset,
etc.) can be selected by the choice of the bulk material parameters and the
strain effects.

Strained GaSb-AlSb,[22] InAs-GaAs,[23] and GaP-AlP[24] have also been grown
and their properties have been investigated.

1.4. Growth methodology

1.4.1. *Growth technology*

The hetero-epitaxial crystal growth technologies have tremendously
advanced in the past two decades. The most critical impact to the ultra-
thin growth techniques was given by molecular beam epitaxy (MBE). The
MBE technology as a means of fabricating *semiconductor superlattice* is
of primary importance. Therefore, we will present this method in this section. We will, in particular, focus on how precise the ultra-thin layers can be
prepared and where problems remain. In addition to the MBE technology,
metal-organic chemical vapor deposition (MO-CVD)[25] has also contributed
to the development of semiconductor superlattices, which are composed of
quarternary III-V compound semiconductors like InGaAsP alloys. More recently, a combination of gas source handling and ultrahigh vacuum (UHV)
technologies such as metal-organic MBE (MO-MBE) or chemical beam epitaxy (CBE)[26], which use metal-organic molecules as source materials, are
becoming popular as well as gas-source MBE[27] using hydride gases like
arsine (AsH$_3$) and phosphine (PH$_3$). This trend occurs because of the possibilities to further control the compositional changes and hetero-interface
qualities and in order to gain useful epitaxial selectivity for the selective
area growth of quantum wires and quantum dots.

Generally speaking MBE is a method of using three temperatures
in binary systems, e.g., the substrate temperature (T_S) and the source
material temperature of the group III (T_{III}) and the group V elements
(T_V) in the case of III-V compounds. Fig. 1.8 schematically shows the
MBE apparatus for the GaAs epitaxial growth. It primarily consists of a

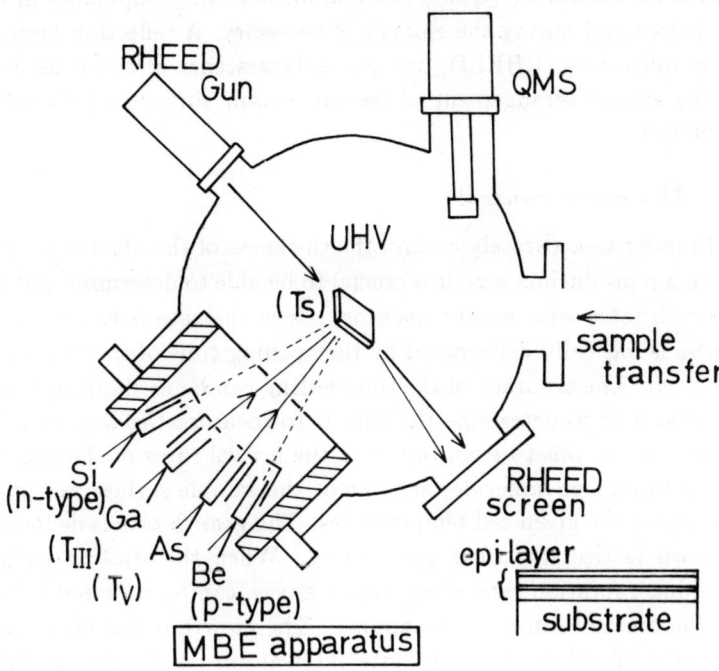

Fig. 1.8. Schematic diagram of the epitaxial growth apparatus for MBE.

UHV chamber, effusion cells for source materials (typically eight sources are controlled independently for group III and V elements and for dopants), a substrate heating holder and a sample transfer tool. The UHV conditions are necessary in order to avoid the incorporation of unwanted impurities so that the pressure should be as low as possible (typically less than $\sim 10^{-8}$ Pa). The three temperatures, T_S, T_{III}, and T_V (and others if necessary) should be separately controlled with an accuracy of $\pm 0.2°C$ or better without any interferences. To avoid thermal interferences and to suppress outgasing from the wall, liquid nitrogen shrouds are often used to surround the substrate heating zone and the cell heaters separately. The sample transfer equipment is used to exchange samples without destroying the UHV and the contamination-free growth conditions, thus, permitting simultaneous growth of a large number of samples in one lot. The quadrupole mass spectrometer (QMS) is useful for detecting impurities in the gas phase before and during the epitaxy, if necessary. A reflection high energy electron diffraction (RHEED) gun and a fluorescence screen is used to monitor the atomic arrangement of the sub-atomic surfaces of the substrate and epilayer.

1.4.2. *Thickness control*

In order to accurately control the thickness of the ultra-thin epitaxial layers in a reproducible way, it is crucial to be able to determine and predict the growth rates with ample precision. Once the rate is known, then the thickness is basically determined by the opening time of the shutter of the effusion cell. The accuracy of the time setting can be as short as 0.1 s. The traditional way to determine the rate is to measure the beam equivalent pressures, which must be converted to the arrival rates of the constituent elements (atoms or molecules arriving at the sample surface per unit area and time) for the given cell temperatures. The growth rate is deduced from the known lattice structural parameters. When the sticking coefficient, i.e., the incorporation rate of impinging fluxes, can be assumed to be constant, the method can easily be applied. The growth of the GaAs layers is mainly determined by the arrival rate of Ga atoms for the arsenic stabilized surfaces under As_4 or As_2 molecule over-pressures. However, the sticking coefficients are generally dependent on the substrate temperature and the surface coverage of the constituent elements. Then, the growth rate is a complicated quantity to be determined, and an empirical method is often employed, which usually gives better results. In this case the most simple but tedious way is just to grow relatively thick, say, 3 μm test layers and measure the thickness by an electron microscope. The growth rate is determined by the thickness divided by the time necessary for the growth under the fixed epitaxy conditions. Another useful empirical way is demonstrated in Fig. 1.9. After growing a set of samples within a short period of time (in order to avoid the change of the growth conditions due to the consumption of the materials), all the dimensional parameters are deduced by X-ray and TEM measurements. In the example shown in Fig. 1.9, four GaAs/AlAs superlattice samples were grown just by changing the shutter opening times for the Al cell without any changes of the other growth parameters. The Ga shutter opening time was fixed at 22 s. The superlattice

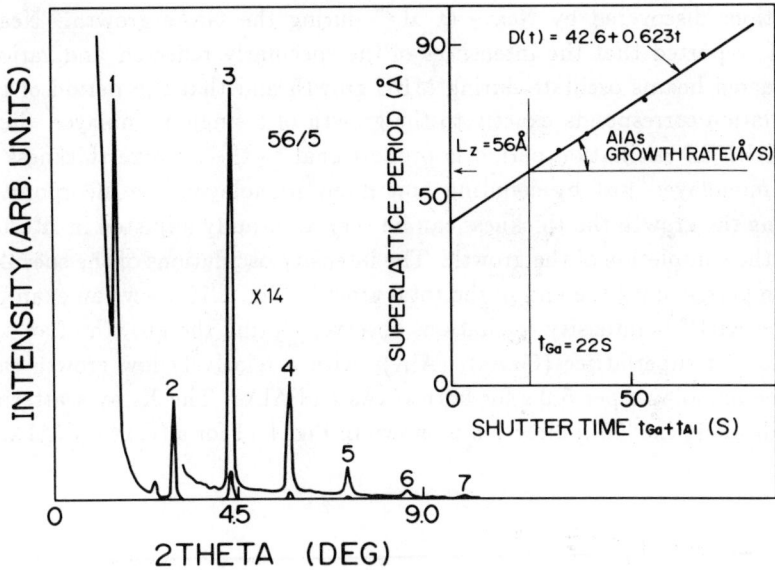

Fig. 1.9. Small-angle X-ray diffraction profile of GaAs/AlAs superlattice ($L_Z = 5.6$ nm, $L_B = 0.5$ nm). The inset shows the relationship between the superlattice period D and the total shutter time t of the Ga and Al cells used for the growth of four superlattice samples. A least squares fit gives $D(t)$, which is used to determine the growth rates of GaAs and AlAs.

period $D = L_Z + L_B$ was measured by small-angle X-ray diffraction experiments in the four samples. The diffraction peaks in the profile of one of the superlattice samples are clearly seen up to the seventh order. In the inset the period D is plotted as a function of the total time $(t_{Ga} + t_{Al})$ of the open shutters for the Ga and Al cells. The AlAs growth rate is obtained from the slope, while the intersection at $t_{Al} = 0$ gives the thickness of GaAs, which is converted into the GaAs growth rate.

1.4.3. *Monolayer thickness control*

Although the above mentioned techniques give practically useful ways of determining the thickness parameters of superlattices, a novel method with higher accuracy was invented by Sano et al.[28] for controlling the ave-

rage thickness of epitaxial layers in multiples of monolayers (called monolayer superlattice). Their method is based on the RHEED intensity oscillations discovered by Neave et al.[29] during the GaAs growth. Neave
et al. reported that the intensities of the specularly reflected and various
diffracted beams oscillate during MBE growth and that the period of the
oscillation corresponds exactly to the growth of a single monolayer. Since
the RHEED oscillation period is proportional to the average thickness of
one monolayer, just by counting how many monolayers are incorporated
during the growth the thickness can be very accurately adjusted in-situ before the completion of the growth. The intensity oscillations of the specular
beam persist until the end of the total growth. Fig. 1.10 shows an example
of the RHEED intensity oscillations observed during the growth of an one
monolayer superlattice $(GaAs)_1$-$(AlA)_1$ with a relatively low growth rate
of one monolayer per 5.0 s for both GaAs and AlAs. The X-ray scattering
profile along the [00L] direction is shown in Fig. 1.11 for a $(GaAs)_2$-$(AlAs)_2$

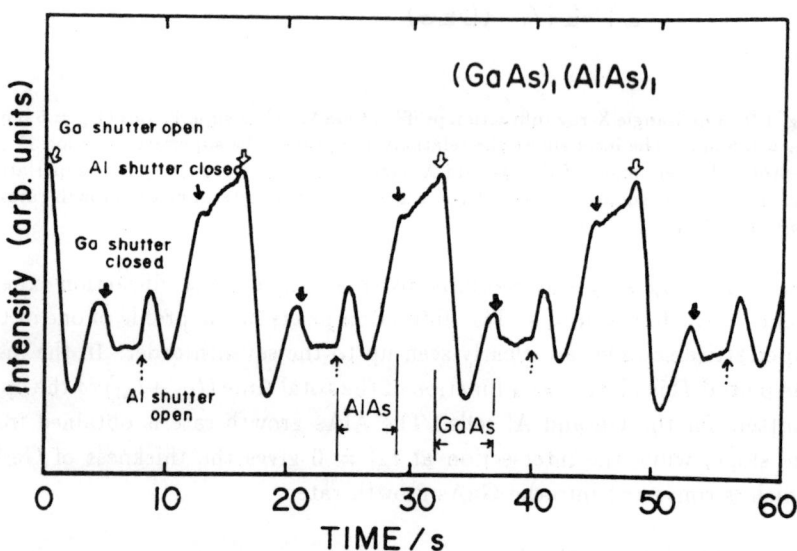

Fig. 1.10. RHEED intensity oscillations of the specular beam during monolayer superlattice growth (from Ref. 28).

superlattice with 300 periods. The unit cell size Λ is two times of the lattice constant (see Fig. 1.1). While the peaks at $L = 2$ and 4 are the fundamental Bragg scattering of the fcc lattice, the superlattice reflections are clearly detected at $L = 0.5, 1.0, 1.5, 2.5, 3.0$ and 3.5. The RHEED intensity oscillation technique is thus useful for the monolayer growth control. However, we should note that these results of monolayer controlled superlattices do not always imply perfect atomic arrangement of the heterostructures as illustrated in Fig. 1.1 and that the interfaces are atomically sharp. Although the arrangement of the majority of the atoms is well ordered in the synthesized superlattices, compositional fluctuations still exist. These effects influence the important physical properties of superlattices such as the absorption spectral linewidth. These issues should be an important subject

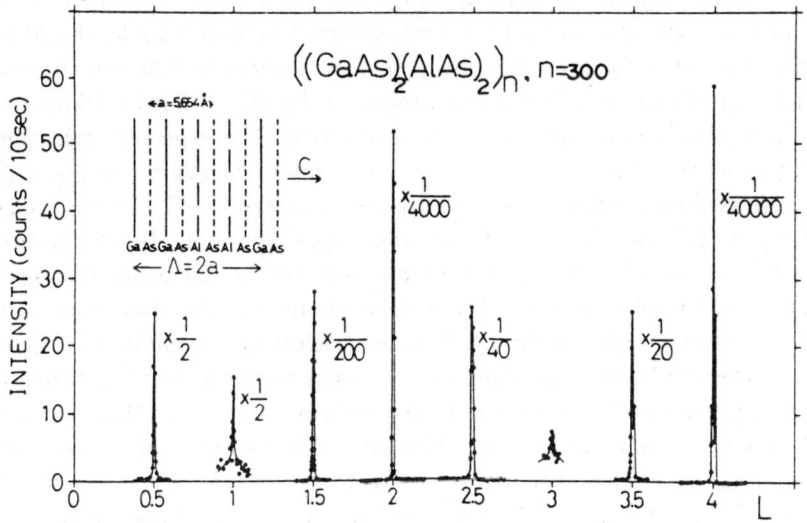

Fig. 1.11. X-ray scattering profile along the [00L] direction and the period $\Lambda = 2a$, $a = 0.5654$ nm. The structure of the grown crystal is schematically shown in the inset (from Ref. 28).

for future growth studies of the atomic-scale heterointerface control.

1.5. Characterization

1.5.1. *Miniband structure*

Structural properties of the superlattices are usually characterized by electron microscope observations and X-ray measurements. These properties have been already discussed in the previous sections in conjunction with the materials and their preparation methodology. In this section we will review their basic electronic properties probed by optical spectroscopy.

In order to optically characterize the superlattice miniband structures, absorption spectra are often employed. In Fig. 1.12 typical examples of the absorption spectral lineshape at room temperature are shown for GaAs/AlAs superlattices near the fundamental aborption edge. These spectra have been obtained using photocurrent spectroscopy.[30] The superlattice layer is embedded in the intrinsic region of a p-i-n diode. A series of four samples #1, #2, #3, and #4 has been designed by only varying the AlAs barrier thickness L_B from 0.57 nm (two monolayers), to 0.86 nm (three), to 1.14 nm (four), to 1.73 nm (six), respectively, with a fixed thickness of 3.13 nm (eleven monolayers) for the GaAs well. The expected transition energies between the electron and heavy-hole miniband edges (see Fig. 1.2) are indicated by horizontal bars, which are calculated based on the Kronig-Penney model assuming the effective mass approximation. Under forward bias ($V_F = 0.6$ V), which nearly corresponds to the flat-band, the spectra marked by solid lines are obtained. We note that the absorption edge shifts to lower energies as the L_B decreases, reflecting the miniband dispersion. When the barrier becomes thin enough, resonant coupling through tunneling processes between the GaAs wells is enhanced. This causes a red-shift of the absorption edge. However, this coupling can be reduced by applying an electric field perpendicular to the superlattice layers, which leads to Wannier-Stark localization (see Chapter 3). The dashed spectra in Fig. 1.12 show the Wannier-Stark localization effect under the reverse bias voltage of 14 V. The miniband dispersion is destroyed and the absorption edge is shifted by the field to higher energies (blue-shift). It should be noted that the development of the quantum well transition (zeroth order Stark ladder transition) occurs in the same wavelength region around 730 nm for all samples. These results give clear evidence of the miniband dispersion.

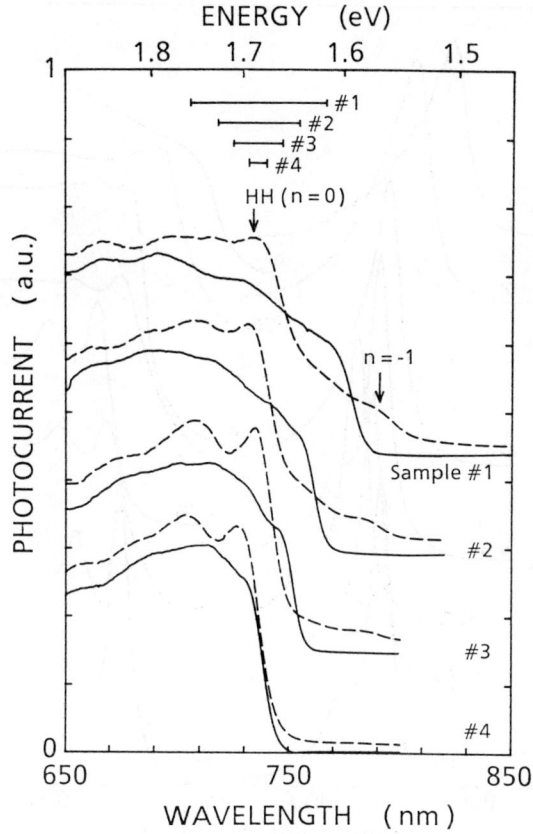

Fig. 1.12. Photocurrent spectra of four superlattice samples #1, #2, #3, and #4 at room temperature (from Ref. 30). Solid curves indicate spectra taken under nearly flat-band conditions and dashed ones under the high electric field. The top horizontal bars correspond to the calculated transition miniband widths. The peak labels denote the Stark-ladder index.

Fig.1.13 shows photoluminescence (PL) and in-plane photocurrent (PC) spectra of a series of seven $(GaAs)_{24}$-$(AlAs)_n$ superlattice samples ($n = 1, 2, 4, 8, 16, 20$, and 36), which were grown using the RHEED intensity oscillation technique discussed in section 1.4.3. These samples were not intentionally doped and the measurements were performed at room tem-

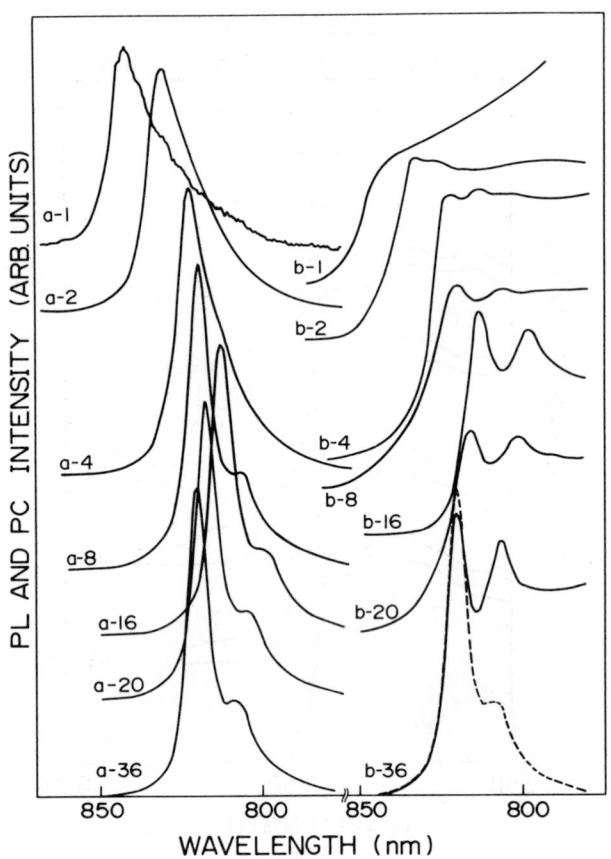

Fig. 1.13. Photoluminescence (PL) and in-plane photocurrent (PC) spectra of $(GaAs)_{24}$-$(AlAs)_n$ monolayer superlattices (the well width is 24 monolayers and the barrier thickness is n monolayers) at room temperature. The n value is given in the PL series (left) by $a - n$, while in the PC series (right) it is labeled by $b - n$. Note that there is no Stokes-shift for the $(GaAs)_{24}$-$(AlAs)_{36}$ superlattice at room temperature when the PC is compared with the superimposed dashed PL spectrum.

perature. As the thickness of the AlAs barriers (n value) decreases, it is clearly seen that the luminescence peak energy as well as the fundamental absorption edge shifts to lower energies. In Fig. 1.14 the PL and PC peak

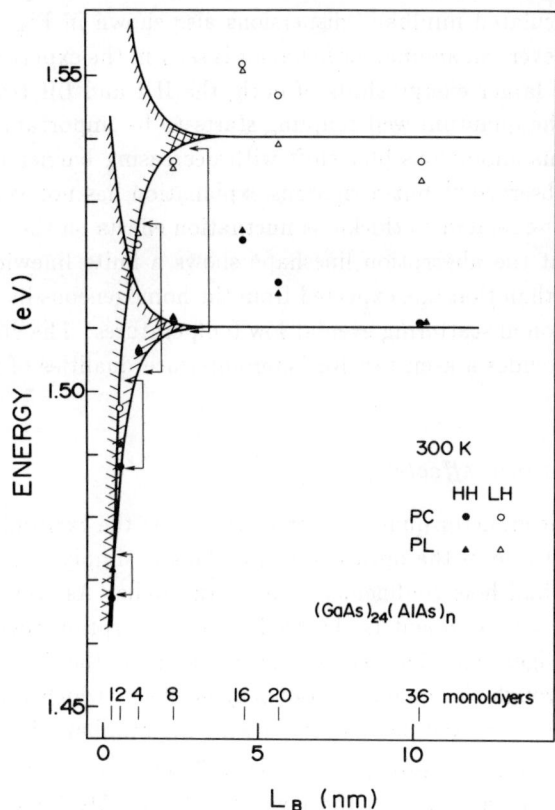

Fig. 1.14. Optical transition energies as a function of barrier thickness for the (GaAs)$_{24}$-(AlAs)$_n$ monolayer superlattices observed in photoluminescence (PL) and photocurrent (PC) spectra at room temperature. The energy ranges of the calculated optical transitions are shown for those related with heavy-hole (HH) and light-hole (LH) subbands by solid curves and hatching. A pair of arrows indicate the calculated LH-HH splitting for $n = 1, 2, 4,$ and 8.

energy positions observed for the heavy-hole (HH) and light-hole (LH) related transitions are plotted. We note that the energy splitting between the HH and LH transitions exhibits a reduction as the AlAs thickness decreases. In the limit of $L_B = 0$, the bulk properties of GaAs should be recovered. The experimental results observed in Fig. 1.14 are generally consistent with the expected behavior and in fact show good agreement with the calculated miniband dispersions also shown in Fig. 1.14 by solid curves. However, an anomalous behavior is seen in the experimental results which shows larger energy shifts of both, the HH and LH transition energies, when the quantum well coupling starts to be important ($n = 16$ and $n = 20$). This anomalous blue-shift with decreasing barrier thickness was previously observed,[31] but a rigorous explanation has not yet been obtained. It may be related to thickness fluctuation effects on the well-coupling. We note that the absorption lineshape shows a finite linewidth, which is much larger than that one expected from the homogeneous linewidth determined by phonon scattering even at low temperatures. The criterion of the linewidth provides a keen test for heterointerface qualities of the quantum structures.

1.5.2. *Excitonic effects*

In semiconductor quantum heterostructures the excitonic effects play an important role in the optical spectra. This is simply due to the stronger electron and hole confinement within the wells. As seen in Fig. 1.13, the excitonic enhancement is observed in the absorption spectra near the absorption edge. The degenerate valence bands at the Γ-point split into subbands through the quantum confinement effect, which leads to heavy-hole and light-hole excitons associated with the miniband bottoms.

In semiconductor superlattices, two different types of excitonic effects should exist. In addition to the lower band-edge M_0 Van-Hove singularity, which is even present in bulk semiconductors, the critical point M_1 at the miniband top should also be observable, which is formed by zone folding along the z-direction. The excitonic effects at this saddle-point were theoretically predicted to be present in superlattices by Chu and Chang.[32] The energy miniband diagram in real space and in k-space is schematically illustrated in Fig. 1.15 for a typical type I superlattice. Fig. 1.16 shows photocurrent spectra of three GaAs/AlAs superlattice p-i-n diodes with (L_Z, L_B) values of (a) (6.26 nm, 0.57 nm), (b) (6.26 nm, 0.86 nm) and (c) (3.92 nm, 0.86 nm) in order to illustrate the wide spectral changes with variations of both L_Z and L_B.[33] The spectra depicted by solid curves were measured near flat-band (under forward bias voltages). In the low temperature PC spectra at 16 K, sharp absorption thresholds are observed.

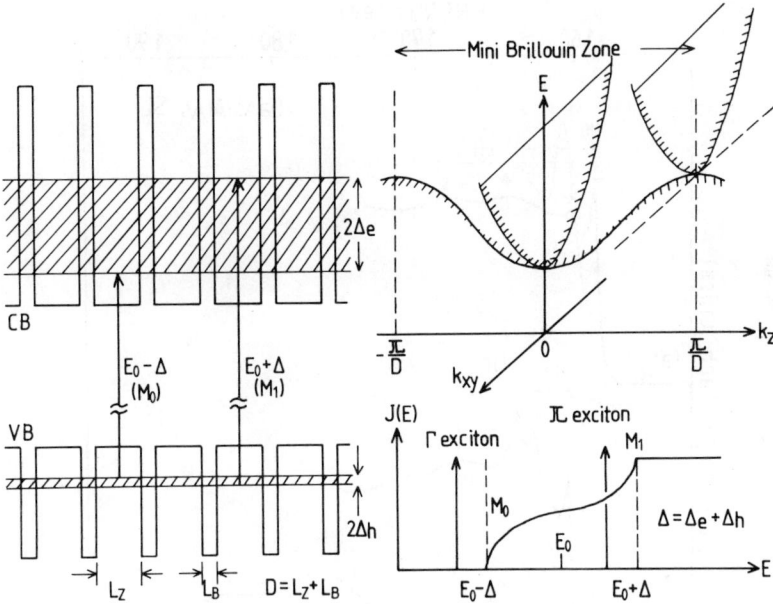

Fig. 1.15. Energy band diagram and miniband dispersion along the z-axis in type I superlattices. At the mini-Brillouin zone center the M_0 type Van-Hove singularity is present in the joint-density of states at the miniband bottom $(E_0 - \Delta)$ by zone folding. At the mini-Brillouin zone edge $(k_z = \pm\pi/D)$ the M_1 saddle-type singular point associated with the upper miniband edge $(E_0 + \Delta)$ appears. E_0 denotes the confinement energy of the quantum well, which is dispersed by tunneling into a miniband of width $2\Delta = 2\Delta_e + 2\Delta_h$. The excitonic transitions Γ and Π associated with such singular points (M_0 and M_1, respectively) are schematically illustrated.

When the AlAs barrier thickness increases, the threshold wavelength (energy) decreases (increases) from the spectrum (a) to (b). At the absorption edge a sharp peak (indicated by an upward arrow) is observed in all spectra, indicating excitonic effects for the optical transitions at the Γ-point. Besides the leading absorption peak, several additional peaks and shoulders are seen in the spectra. At first glance a second sharp peak is observed next to the leading peak, which is indicated by a left downward arrow. We note that the energy separation between these two neighboring peaks increases with L_B. This observation can be explained in the same way as discussed in Figs. 1.13 and 1.14. Broader peaks indicated by arrows on the right-hand

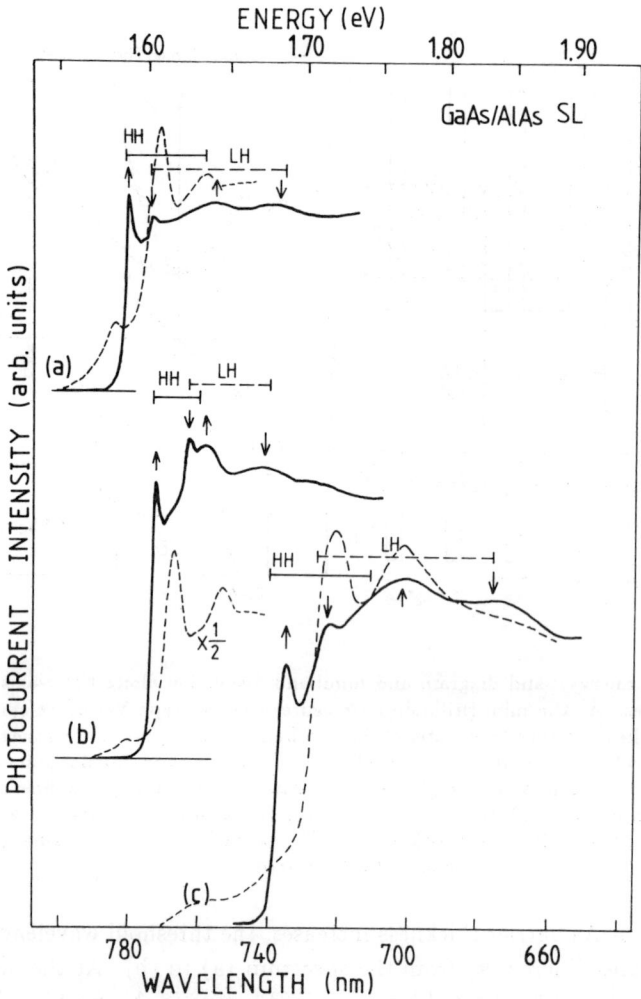

Fig. 1.16. Low temperature photocurrent spectra of three GaAs/AlAs superlattice diode samples with: (a) $L_Z = 6.26$ nm, $L_B = 0.57$ nm, (b) $L_Z = 6.26$ nm, $L_B = 0.86$ nm, and (c) $L_Z = 3.92$ nm, $L_B = 0.86$ nm. The solid curves represent spectra measured near flat-band (positive bias from $+1.1$ to $+1.4$ V), while the dashed ones were measured under perpendicular electric fields (reverse bias at -3 V for (a) and (b) and at -10 V for (c)). The calculated wavelength ranges for the superlattice heavy-hole (light-hole) transitions are indicated by horizontal solid (dashed) bars. Upward (downward) arrows indicate prominent transitions related to heavy-hole (light-hole) excitons.

side are located at shorter wavelengths (higher energies) and their transition energies decrease with increasing L_B. When the well width is decreased from (b) to (c), keeping the barrier thickness constant, the threshold peak is blue-shifted and broader peaks are observed with larger energy separations between them.

The experimental results depicted in Fig. 1.16 can be explained rigorously by miniband dispersion effects. The band width of the miniband formed along the z-axis is varied by choosing the L_Z and L_B values. In this case we have Van-Hove singularities of the M_0 type at the miniband bottom and of M_1 type at the upper miniband edge. When L_B is decreased, the edge of the fundamental optical absorption should be red-shifted. The experimental observation in Fig. 1.16 exactly corresponds to this expected behavior. The high-energy structures observed in the spectra are attributed to excitonic effects which are associated with the M_1 Van-Hove singularity (II-excitons). A peak nature of the optical absorption indicates excitonic effects as the origin of the HH-related as well as LH-related transitions. It is noted that the peaks of the high-energy II- and low energy Γ-excitons are in excellent agreement with the calculated miniband widths.

When an external electric field is applied along the z-direction, however, additional fine structure appears as shown in Fig. 1.16 by the dashed curves. We note that in the dashed curves a main steep edge always appears at longer wavelengths of the threshold peak of the corresponding flat-band spectra. At the same time, the energy separation between them increases with decreasing L_Z and L_B. The results of the external field effects are well explained in terms of Wannier-Stark localization. The weaker edge at longer wavelengths observed in the dashed spectra of Fig. 1.16 is attributed to the spatially indirect first-order Stark ladder transition. When the applied electric field is large enough to generate zeroth order Stark-ladder, i.e., interwell direct, transitions, they should be located near the center of the optical-transition miniband width. The experimental observations in Fig. 1.16 directly prove this for both, HH and LH transitions. A careful comparison of the solid and dashed spectra reveals that the line-shape of the superlattice Γ-excitons is much sharper than that of quantum well excitons in all three samples. The linewidth narrowing of the superlattice Γ-excitons is also ascribed to miniband dispersion effects, which average out the interface potential fluctuations over the length scale of the superlattice exciton wavefunction along the z-direction. Thus, the energy position and the linewidth of the optical transitions are all consistent with miniband dispersion effects in the fundamental absorption spectra.

References

1. L. Esaki and R. Tsu, *IBM J. Res. Dev.* **4**, 61 (1970).
2. R. Dingle, W. Wiegmann, and C.H. Henry, *Phys. Rev. Lett.* **33**, 827 (1974); R. Dingle, in *Festkörperprobleme, Vol.XV*, edited by H.J. Queisser (Pergamon Vieweg, Braunschweig, 1975) p.21.
3. A.Y. Cho and J.R. Arthur, *Prog. Solid State Chem.* **10**, 157 (1975).
4. Y. Horikoshi, *Semicond. Sci. Technol.* **8**, 1032 (1993).
5. L. Däweritz and K. Ploog, *Semicond. Sci. Technol.* **9**, 123 (1994).
6. E.O. Göbel and K. Ploog, *Prog. Quant. Electr.* **14**, 289 (1990); K. Ploog, *Angew. Chem. Int. Ed. Engl.* **27**, 593 (1988); *Cryst. Prop. Prep.* **31**, 111 (1991).
7. E.H.C. Parker (editor), *The Technology and Physics of Molecular Beam Epitaxy* (Plenum Press, New York, 1985).
8. K. Ploog and G. Döhler, *Adv. Phys.* **32**, 285 (1983).
9. R.C. Miller, D.A. Kleinmann, and A.C. Gossard, *Phys. Rev. B* **29**, 7085 (1984); J.M. Langer and H. Heinrich, *Phys. Rev. Lett.* **55**, 1414 (1985).
10. K. Fujiwara, R. Cingolani, and K. Ploog, *Solid State Commun.* **72**, 389 (1989).
11. L. Esaki, *IEEE J. Quantum Electron.* **QE-22**, 611 (1986).
12. P. Dawson, B.A. Wilson, C.W. Tu, and R.C. Miller, *Appl. Phys. Lett.* **48**, 541 (1986).
13. R. Cingolani, L. Baldassarre, M. Ferrara, and K. Ploog, *Phys. Rev. B* **40**, 6101 (1989).
14. K. Kawashima, K. Fujiwara, and K. Katahama, *Superlattices Microstruct.* **7**, 331 (1990); H. Schneider, K. Kawashima, and K. Fujiwara, *Phys. Rev. B* **44**, 5943 (1991).
15. R. Merlin, K. Bajema, R. Clarke, F.-Y. Juang, and P.K. Bhattacharya, *Phys. Rev. Lett.* **55**, 1768 (1985).
16. G.T. Einevoll and L.J. Sham, *Phys. Rev. B* **46**, 7787 (1992).
17. J. Faist, F. Capasso, D.L. Sivco, C. Sirtori, A.L. Hutchinson, and A.Y.Cho, *Science* **264**, 553 (1994).
18. Y. Shiraki and S. Fukatsu, in *Proceedings of the 22nd International Conference on the Physics of Semiconductors*, edited by D.J. Lockwood (World Scientific, Singapore, 1994), in press.
19. A.V. Nurmikko, in *Proceedings of the 22nd International Conference on the Physics of Semiconductors*, edited by D.J. Lockwood (World Scientific, Singapore, 1994), in press.
20. A.Y. Cho, *Appl. Phys. Lett.* **19**, 467 (1971).
21. A. Gomyo, K. Makita, I. Hino, and T. Suzuki, *Phys. Rev. Lett.* **72**, 673 (1994).

22. P. Voisin, in *Two-Dimensional Systems, Heterostructures and Superlattices* edited by G. Bauer, F. Kuchar and H. Heinrich (Springer-Verlag, Berlin, 1984) p. 192.

23. T. Fukui and H. Saito, in *Inst. Phys. Conf. Ser. 79* (Adam Hilger, Bristol and Boston, 1986) p. 397.

24. K. Asami, H. Asahi, T. Watanabe, S. Gonda, H. Okumura, and S. Yoshida, *Surf. Sci.* **267**, 450 (1992).

25. P.D. Dapkus, H.M. Manasevit, K.L. Hess, T.S. Low, and G.E. Stillman, *J. Cryst. Growth* **55**, 10 (1981).

26. W.T. Tsang, *J. Cryst. Growth* **81**, 261 (1987).

27. M.B. Panish, *J. Cryst. Growth* **81**, 249 (1987).

28. N. Sano, H. Kato, M. Nakayama, S. Chika, and H. Terauchi, *Jpn. J. Appl. Phys.* **23**, L640 (1984).

29. J.H. Neave, B.A. Joyce, P.J. Dobson, and N. Norton, *Appl. Phys. A* **30**, 1 (1983).

30. K. Kawashima, K. Fujiwara, T. Yamamoto, M. Sigeta, and K. Kobayashi, *Jpn. J. Appl. Phys.* **30**, L793 (1991).

31. J.L. de Miguel, K. Fujiwara, L. Tapfer, and K. Ploog, *Appl. Phys. Lett.* **47**, 836 (1985).

32. H. Chu and Y.-C. Chang, *Phys. Rev. B* **36**, 2946 (1987).

33. K. Fujiwara, K. Kawashima, T. Yamamoto, N. Sano, R. Cingolani, H.T. Grahn and K. Ploog, *Phys. Rev. B* **49**, 1809 (1994).

CHAPTER 2

MINIBAND TRANSPORT

by ALAIN SIBILLE

2.1. Introduction: Miniband transport, dream or reality?

Early interest in minibands arose when physicists became aware of the unusual properties exhibited by narrow band solids in contrast to the wide bands of natural crystalline materials.[1,2] An important step was achieved when it was realized by Esaki and Tsu in 1970 that such minibands could practically be obtained in man-made structures through modern semiconductor epitaxial techniques.[3] The very simple, but revolutionary, idea was to mimic nature by growing periodically spaced alternated layers of different semiconductors, thus producing artificial crystals of relatively large effective lattice constant in one direction. From then on several attempts worldwide intended to demonstrate miniband effects in these 'superlattices' (SL). However, the basic phenomenon at the origin of the very existence of minibands is *tunneling*. Because of the exponential decay of the electron or hole wavefunctions with distance, tunneling requires extremely thin semiconductor layers in order to obtain appreciable miniband widths. Therefore, the formidable task represented by the ability to grow reproducibly and in a controlled manner multilayers of nanometer-sized periods has been a major challenge for nearly two decades. The latter has stimulated considerable research aiming at fabricating superlattices and other quantum structures in order to determine if miniband transport was a dream or reality, and the result was a spectacular progress in growth.

The successful achievement of quantum well devices has preceded that of real superlattices[4] because of less stringent requirements on layer sizes.

In the GaAs/AlGaAs material system, quantum wells as thick as about 10 nm allow to measure easily quantum confinement energies on the order of a few tens of meV by standard optical spectroscopy techniques. However, much thinner barriers are necessary in order to obtain significantly wide minibands. A related research area is concerned with double barrier tunnel devices, which were also proposed and demonstrated by Esaki and Tsu as early as 1974 in order to produce negative differential resistance devices.[5] The added complexity of the superlattice, however, for which a periodicity of the multilayer structure must also be achieved, demands much more from the crystal grower. This profound technological difficulty has impeded the conclusive experimental observation of miniband transport effects in semiconductor superlattices until the very recent years. It is the purpose of this chapter to describe some aspects of miniband transport and to review a few of the important contributions to this field, which opened the way to make the dream become a reality.

2.2. Miniband structure: Calculation and experimental observation

2.2.1. Generalities on superlattice minibands

The schematic structure of a periodic superlattice is shown in Fig. 2.1, where A and B are two different semiconductor materials of respective layer thicknesses a and b (period: $d = a + b$). In principle such an arrangement defines a new crystal whose band-structure can be directly determined from the atomic potentials of the various constituent atoms, with reduced minibands, minigaps, and minizones. However, when a and b are not too small compared with the interatomic spacing, an adequate approximation is obtained by replacing these fast varying potentials by an effective potential derived from the band-structure of the original bulk semiconductors. In a one-dimensional effective mass approximation along z, this crenel-shaped potential is depicted in Fig. 2.1. It is straightforward to solve the 1D Schrödinger equation in each of the individual layers, whose solutions ψ are linear combinations of real or imaginary exponentials. However, the boundary conditions at the interfaces, where the potential varies rapidly, are not really trivial and have been the subject of many theoretical studies tending to justify some commonly employed connection rules between the wavefunctions. It is generally assumed that the wavefunction is continuous.

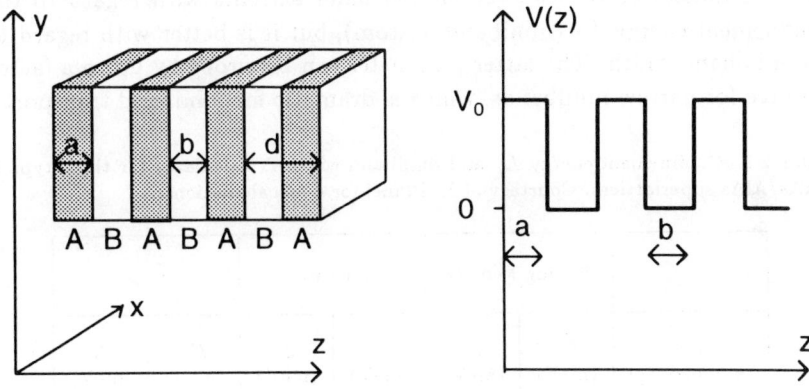

Fig. 2.1. Schematic structure of a superlattice composed of materials A and B grown along the z axis (left). The effective potential experienced by electrons is shown on the right.

The historical *Kronig-Penney model* also assumes the continuity of the wavefunction derivative, and takes a weighted average over the two materials for the unique effective mass. Slightly more refined is a model in which the difference of effective mass m^* between the two materials is taken into account, with a connection rule specifying the continuity of $m^{*-1} \, d\psi/dz$ (Ref. 6). However, this model requires an educated choice for the effective masses, since the wavefunctions depend sensitively on m^*. If the energy level of the state under consideration lies high in the conduction band of, e.g., material A, nonparabolicity will have to be accounted for in the choice of the effective mass for this layer. Conversely, if the energy level lies deep in the bandgap of material B, the effective mass to be used is a tunneling mass, which will be smaller than the mass of the conduction band bottom. These differences may appreciably affect the calculated positions and widths of the minibands. A full $\vec{k} \cdot \vec{p}$ calculation based on the Kane model helps to correct the effective masses and to obtain more reliable quantities. We show in Tab. 2.1 an example of these differences tested in the GaAs/AlAs system. The conclusion for this particular material combination is clear with respect to the assumed more accurate results based on the $\vec{k} \cdot \vec{p}$ model: the one mass Kronig-Penney model is worse than the model employing effective masses at the two conduction band extrema with regard to the confinement energy E_1 (miniband bottom), but it is better with regard to the miniband width. The latter parameter can be wrong by up to a factor of three for narrow minibands, which is dramatic for miniband transport.

Table 2.1. Confinement energy E_1 and miniband width Δ calculated for three typical GaAs/AlAs superlattices. Courtesy of R. Planel for $\vec{k} \cdot \vec{p}$ calculations.

	Kronig-Penney		Bastard		$\vec{k} \cdot \vec{p}$	
a/b (monolayers)	E_1 (meV)	Δ (meV)	E_1 (meV)	Δ (meV)	E_1 (meV)	Δ (meV)
20/10	81.5	1.6	90.2	0.8	102.8	2.3
10/5	198.1	67.8	202.6	49.3	224.3	72.4
5/5	388.1	198.7	391.5	152	428.7	194.6

Obviously, the essential features of textbook quantum mechanics are verified in superlattice minibands, e.g., the fast lowering of the miniband position, when the well size is increased, or the broadening of the miniband by the tunneling effect, when the barrier becomes thinner. For a large barrier thickness, tunneling is a weak perturbation with regard to the uncoupled dispersionless states, which are fully confined in the well. In this case the dispersion relation $E_z(k_z)$, periodic over $2\pi/d$ with $d = a + b$ by virtue of the Bloch theorem, is fully sinusoidal

$$E_z(k_z) = \frac{\Delta}{2} \left[1 - cos(k_z \, d) \right], \tag{2.1}$$

and the effective mass

$$m^* = \frac{\hbar^2}{\partial^2 E/\partial k^2} \bigg|_{k=0} \tag{2.2}$$

changes sign for $k_z = \pi/(2d)$. In the case of wider minibands this sinusoidal character is no longer preserved as illustrated in Fig. 2.2. For very wide minibands, e.g., when the barrier is so thin that the SL barely exists, the dispersion at the miniband bottom is very close to that of the original well material. Only high up in the miniband for wavevectors well beyond $\pi/(2d)$, is the top actually 'sensed' and does the effective mass change sign. The very shape of the miniband dispersion obviously influences miniband transport profoundly, and accurate dispersion relation calculations are indispensable in the case of wide minibands. It is worth mentioning here that the condition for observing *single miniband transport* is the absence of any interminiband transfer by any process. This means in particular that the thermal quantum $k_B T$ should be much smaller than the energy difference

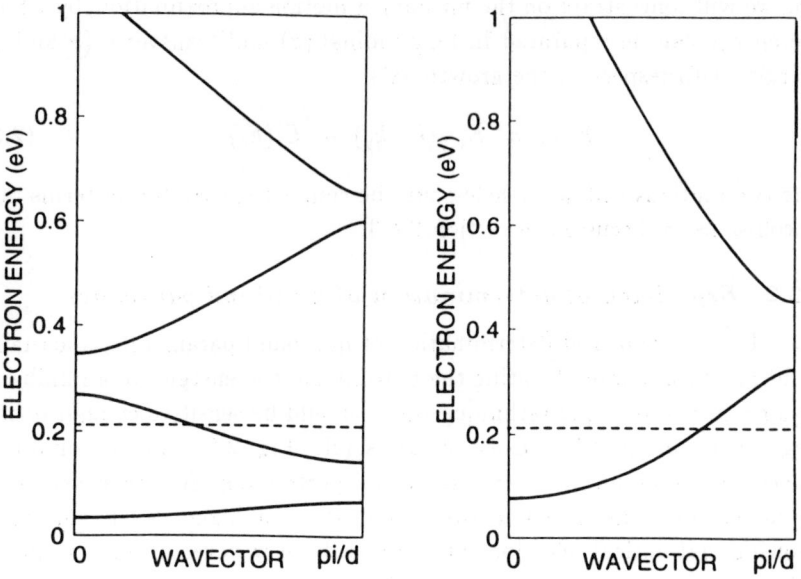

Fig. 2.2. Dispersion relation of two GaAs/AlGaAs superlattices of Al composition 30% and well/barrier widths of 7.0/2.0 nm (left) and 2.5/1.5 nm (right) (the dashed line depicts the top of the barrier). The ground state miniband is clearly not sinusoidal in the latter case.

$E_2 - E_1$ between the first and second miniband, *even in the presence of the applied electric field.* As will be discussed in section 2.6.2 the electron temperature may indeed largely exceed the lattice temperature. This also means that interminiband tunneling is neglected. This effect has been calculated theoretically,[7] although it has not yet been demonstrated experimentally. Briefly stated, interminiband tunneling occurs when the energy levels of two Wannier-Stark states (see Chapter 3 of this volume) originating from different minibands and localized on reasonably close superlattice periods, align at a particular value of the electric field. This condition may only be realized if the minigap is small enough or the electric field very high. Consequently, it is necessary in superlattices designed for single miniband transport to push the second miniband high enough so that these effects can be neglected.

Real superlattices are actually three-dimensional, which means that the energy dispersion $E(\vec{k})$ depends on the *vector* $\vec{k}$. In the following discussion we will concentrate on the uncoupled motion approximation, by which the energy can be separated in longitudinal (z) and transverse (x and y) energies with respect to the growth axis

$$E(\vec{k}) \; = \; E_{x,y}(k_x, k_y) \; + \; E_z(k_z) \, . \tag{2.3}$$

The two motions will nevertheless still be coupled via scattering terms due to collisions as discussed in section 2.4.3.

2.2.2. *Experimental determination of miniband parameters*

The experimental determination of miniband parameters is actually a difficult task: simply locating the bottom and top energies of a miniband requires a spectroscopic technique which should be sensitive enough to the singularities of the SL density of states (cf. Fig. 2.3). In photoluminescence and photoluminescence excitation spectroscopy, it is relatively easy to observe excitons linked to the ground-state or higher level minibands knowing that optical selection rules impose both wavevector and miniband index conservation in the transition. The most prominent features in the optical spectra of direct (type I) superlattices are due to $k = 0$ excitons involving either light-hole or heavy-hole states. However, they provide no direct information on miniband widths, since, for instance, the observation of $k = \pi/d$ excitons is also needed. Such *saddle point excitons* have smaller oscillator strengths than zone center excitons and are therefore more

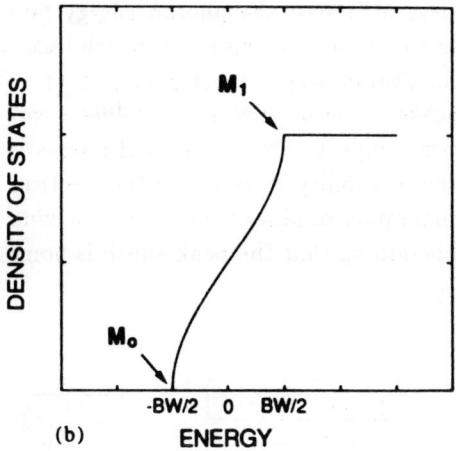

Fig. 2.3. Singularities at the miniband extrema of the density of states in a superlattice. Here BW refers to the miniband width (from Ref. 8).

difficult to observe unambiguously in optical spectra. Several reports of such saddle-point excitons have indeed confirmed simple calculations of the miniband widths using effective mass approximations or tight-binding models.[8–12]

One different method relies on the application of a magnetic field perpendicular to the growth axis. Landau orbits in this configuration are sensitive to both transport in the layer plane and across the SL barriers, which provides an opportunity for determining some miniband parameters such as the electron effective mass along z. This was first done by Duffield et al.[13] in GaAs/AlGaAs SLs through cyclotron resonance experiments, who measured electron effective masses in fair agreement with calculations based on the envelope-function approximation. More recently, a saturation of the cyclotron energy with increasing magnetic field could successfully be associated with the miniband top.[14] Finally, a related but more informative approach was followed by Maan and co-workers by magneto-optical experiments.[15] The key point in these experiments is the dispersive character of the electron or hole state in the Landau gauge. When their total energy lies *outside of the electron or hole miniband*, it depends periodically on the position of the cyclotron orbit center along the SL axis. Allowed interband Landau level optical transitions are therefore dramatically broadened by this dispersion, unless the photon energy falls within the combined electron plus hole miniband width, in which case well-defined peaks are seen in the photoluminescence spectra (Fig. 2.4). The energy range, over which these peaks are seen, provides therefore a way to access the miniband widths experimentally. One additional interesting consequence of this technique is the possibility to estimate the electron coherence length from the maximum number of observable peaks knowing the extent of the Landau levels and assuming that the peak width is homogeneously related

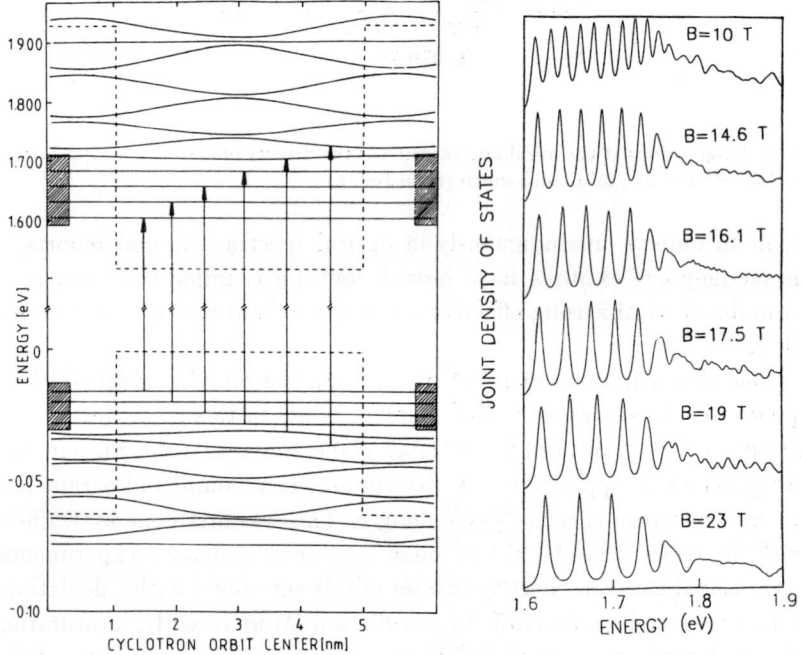

Fig. 2.4. Left: Landau levels vs cyclotron orbit center under a magnetic field applied parallel to the layers in a superlattice. The dashed areas depict the electron and hole minibands, and the arrows indicate the spectral resonances in optical spectroscopy. Right: Calculated joint density of states under various magnetic fields (from Ref. 15).

to the electron state natural lifetime. Since the electron coherence length is related to the SL 'perfection', it plays an important role in miniband transport.

One drawback of interband optical methods is the involvement of *both electrons and holes*, which doubles the difficulty in the extraction of significant data. Particularly noteworthy is the implication of excitonic effects which shift optical transition energy positions by the amount of the exciton binding energy. In wide enough bandgap materials with appreciable electron effective masses, shifts as high as 10 meV are not rare. This is of course not negligible in narrow miniband SLs taking into account the uncertainty in exciton binding energy. This problem can be avoided by using *interminiband absorption*, which allows to measure the sum of the ground and first excited electron miniband widths and to extract the joint density of states for these two minibands. In principle it can be applied to any couple of minibands for which there is no strict forbidden selection rule, but the oscillator strength rapidly decreases with the miniband index. One other drawback is the high density of electrons required for measuring an absorption at all, which can affect the measurements through many-body effects. Experimental spectra are nevertheless in quite a good agreement with experiments taking into account the Fermi-Dirac distribution in the miniband and additional impurity transitions (cf. Fig. 2.5).[16]

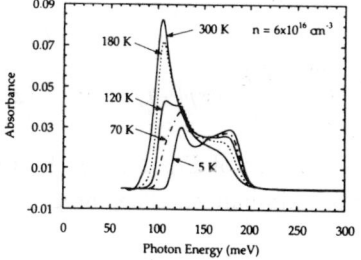 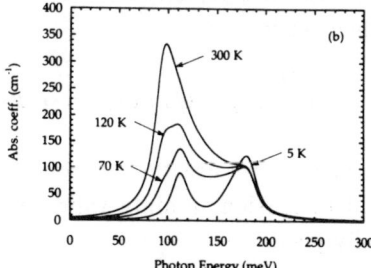

Fig. 2.5. Experimental (left) and theoretical (right) interminiband optical absorption spectra (from Ref. 16).

2.3. Semiclassical miniband transport models

2.3.1. *Esaki-Tsu model*

The fundamental effect of the existence of a miniband in electron or hole conduction is related to an upper limit of the energy, which can be acquired from the electric field applied along the SL growth axis. As opposed to other saturation mechanisms commonly known in bulk semiconductors, this limit should strongly influence high field transport for moderate miniband widths Δ. We summarize here the seminal paper by Esaki and Tsu,[3] whose essential features have been demonstrated to be well founded in the last few years. We begin with the semiclassical equation of electron motion in a one-dimensional solid, to which an electric field F is applied

$$\hbar \frac{dk}{dt} = e\,F \quad \text{and} \quad v_g = \frac{1}{\hbar}\frac{\partial E}{\partial k}\, , \qquad (2.4)$$

where v_g denotes the group velocity. The first equation predicts a linear variation of the electron wavevector with time $k = k(0) + eFt/\hbar$, which translates into a *periodic* time dependence of the electron group velocity. This conclusion originates from the periodicity $2\pi/d$ of the energy dispersion in k-space.

However, these time dependences neglect scattering and are therefore unrealistic. In a phenomenological approach, scattering characterized by a mean collision time τ was introduced through an assumed exponential temporal decay of the probability of collisionless (ballistic) transport at time t. The net velocity computed as an average over the whole electron gas then reads

$$V = \int_0^\infty exp(-t/\tau)\,dv_g = e\,F\,\hbar^{-2}\int_0^\infty \frac{\partial^2 E}{\partial k^2}\,exp(-t/\tau)\,dt\ . \quad (2.5)$$

From now on we use the sinusoidal approximation of Eq. 2.1 for the miniband dispersion, mainly valid for narrow minibands. The integration trivially yields

$$V = \frac{\mu\,F}{1 + (F/F_c)^2} \quad \text{with} \quad \mu = \frac{e\,\Delta\,\tau\,d^2}{2\,\hbar^2} \quad \text{and} \quad F_c = \frac{\hbar}{e\,\tau\,d}\ . \quad (2.6)$$

We recognize in these expressions dependences which are well-known to solid-state physicists: firstly, the *low-field mobility* μ is directly proportional to the scattering time; secondly, μ can also be written as $\mu = e\tau/m^*$, where m^* denotes the electron effective mass at the bottom of the miniband given by $2\hbar^2/(\Delta d^2)$ for narrow minibands. The dependence on Δ originates from that of m^* due to the inverse proportionality between the energy dispersion curvature at $k = 0$ and Δ.

This model therefore predicts negative differential velocity (NDV), i.e., the decrease of the electron mean velocity with increasing electric field beyond a critical field F_c given above. This sole feature has triggered a number of both theoretical and experimental investigations since 1970 and is at the heart of the present discussion.

Let us first remark that NDV here is indeed a true miniband effect, since the peak velocity V_p is given by $V_p = \Delta d/(4\hbar)$, i.e., a value directly proportional to the miniband width. In other words, a very wide miniband, e.g., 2 eV as in a natural solids, would have a peak velocity calculated in this manner orders of magnitude too high to be sensible, and other saturation mechanisms like the intervalley scattering will prevail over miniband NDV.

It is interesting to understand the basic origin of the NDV in this simple model. Actually the absence of scattering would make the electron wavevector grow indefinitely, and its group velocity experience a periodic **Bloch oscillation**. Scattering is somewhat expected to inhibit Bloch oscillations, but it still allows electrons to be accelerated high enough in the miniband, if the electric field is sufficiently large. More precisely, we directly deduce from above the mean electron momentum

$$< k > = \frac{1}{\tau} \int_0^\infty exp(-t/\tau) \, k(t) \, dt = \frac{e \, F \, \tau}{\hbar} , \qquad (2.7)$$

i.e., $< k >= 1/d$ for $F = F_c$. This implies that the average momentum acquired by the acceleration is only $1/d$ at the onset of NDV, i.e., *well below* $2\pi/d$, *which is the momentum acquired by the electron after one full Bloch oscillation*. However, $< k >$ is much closer to the center of the Brillouin minizone located at $k = \pi/2d$, where the dispersion relation has an inflection point and the effective mass changes sign. We can therefore state that in this model the basic origin of NDV is the **negative effective mass** of electrons accelerated up to the energy of the center of the miniband. This conclusion will be later confirmed on the basis of experimental results supported by more sophisticated models. Finally, it is worth mentioning that, although the previous predictions originate from the sinusoidal dispersion relation assumed above, they still remain valid in the case of less simple dispersion relations.[3]

2.3.2. *Relaxation time approximation*

The Esaki-Tsu model as desribed above is actually oversimplified, because

i) it is one-dimensional,

ii) it assumes that a collision fully annihilates the electron momentum, i.e., $k = 0$ after the collision,

iii) it reduces the initial distribution of electron momentum to a Dirac δ-function in wavevector space, i.e., it neglects the thermal broadening of this distribution at finite temperatures,

iv) it does not take into account the various scattering mechanisms, which are important in a real solid, and it only specifies them by a single and constant collision time τ.

In spite of all these shortcomings and drawbacks, the Esaki-Tsu model is surprisingly pertinent, at least qualitatively, as will be shown in the next sections. However, it is necessary to incorporate more realistic ingredients into the theory in order to have a chance of obtaining a quantitative agreement with experiment and of extracting significant information on the physical processes involved in the miniband transport.

The Boltzmann equation provides a natural frame to these expectations particularly by surmounting simplifications i) to ii). In the case of a homogeneous infinite medium without a magnetic field, the distribution function f of variables t and $\vec{k}$ is determined by

$$\frac{\partial f}{\partial t} + \frac{e\,F}{\hbar}\,\frac{\partial f}{\partial k_z} \;=\; S(f) \,-\, \frac{f}{\tau}\,, \tag{2.8}$$

where the electric field is along z and the scattering-in and scattering-out collision operators are of the form

$$S(f) \;=\; \sum_{\vec{k}'} W(\vec{k}', \vec{k})\, f(\vec{k}') \quad \text{and} \quad \frac{1}{\tau} \;=\; \sum_{\vec{k}'} W(\vec{k}, \vec{k}')\,. \tag{2.9}$$

At first we can consider a simplified solution of this equation in the well-known *relaxation time approximation* by virtue of which the scattering-out term is simplified to $S(f) = f_0/\tau$. Here $f_0 = A\, exp[-E/(k_B T)]$ refers to the equilibrium Maxwell-Boltzmann distribution in a non-degenerate SL with E denoting the total kinetic energy from Eq. 2.3.

It turns out that the Boltzmann equation is now separable. In the case of a stationary solution, $f(\vec{k}) = f_{xy}(k_x, k_y) f_z(k_z)$ with

$$\frac{e\,F}{\hbar} \frac{\partial f_z}{\partial k_z} = \frac{f_{0,z} - f_z}{\tau} \quad \text{and} \quad f_{xy} = f_{0,xy} \,, \tag{2.10}$$

where $f_{0,z}(k_z) = B \, exp[-E_z(k_z)/(k_B T)]$ and B is a normalization constant. $f_{0,xy}(k_x, k_y) = A \, exp[-\hbar^2(k_x^2 + k_y^2)/(2m^* k_B T)]$ with A being a normalization constant. This equation can easily be solved by a Fourier transform of the k_z-dependence[17] taking advantage of the $2\pi/d$ periodicity in k_z

$$f(k_z) = \sum_n f^{(n)} \, exp(i\,n\,k_z\,d) \,, \tag{2.11}$$

yielding the immediately soluble equation

$$\left(i\,n\,d\,F - \frac{\hbar}{e\,\tau}\right) f^{(n)} = -\frac{\hbar}{e\,\tau} f_0^{(n)} \,. \tag{2.12}$$

The model is therefore effectively still one-dimensional, but it accounts for the thermal broadening due to the finite lattice temperature through the equilibrium distribution f_0. The drift velocity for a sinusoidal miniband averaged over all wavevectors reads after some algebra

$$V = \frac{\displaystyle\int_{-\pi/d}^{\pi/d} f_z(k_z)\, v_g(k_z)\, dk_z}{\displaystyle\int_{-\pi/d}^{\pi/d} f_z(k_z)\, dk_z} = \frac{i\,\Delta\,d}{4\,\hbar} \frac{f^{(1)} - f^{(-1)}}{f^{(0)}} \tag{2.13}$$

$$= \frac{I_1(\Delta/(2k_B T))}{I_0(\Delta/(2k_B T))} \frac{\mu\,F}{1 + (F/F_c)^2} \tag{2.14}$$

with I_1 and I_0 denoting Bessel functions of second kind, and μ and F_c the same low-field mobility and critical field, respectively, as above.

The relaxation approximation is therefore equivalent to the Esaki-Tsu model except for an additional temperature dependence only entering through the I_1/I_0 factor. The latter is equal to 1 when $k_B T \ll \Delta/2$, and asymptotically decreases as $1/T$ at elevated temperatures. The physical origin of this *thermal saturation of miniband transport* is clear: under low electric field it directly originates from the finite population of non-zero momentum states according to the Maxwell-Boltzmann statistics. Such states contribute to the net conductivity in proportion of their inverse effective mass (cf. Eq. 2.2) along z, which decreases (and eventually becomes negative) with increasing values of k_z. Under a high electric field, we can still understand that carriers have kept the 'memory' of their initial momentum so that the distribution function broadening is at the image of its equilibrium value.

This thermal saturation was first experimentally demonstrated by Brozak et al.[18] through far-infrared (FIR) Drude optical absorption investigations in narrow miniband GaAs/AlGaAs superlattices (Δ=3 meV). This technique requires that the electric field of the radiation is oriented along the SL growth axis in order to measure miniband transport and not in-layer conduction. To maximize the coupling between the FIR light and the SL, a grating was deposited on the sample surface. Finally, a high magnetic field along z permitted to get rid of any parallel transport contribution (an additional important precaution justified by the much larger mobilities of in-plane conduction) due to the extreme quantum limit responsible for full electron localization along x, y. We reproduce below the temperature dependence of the conductivity, which decays with T as expected from the simple model above. It is actually rather surprising that the temperature dependence predicted by miniband conduction was observed in a miniband

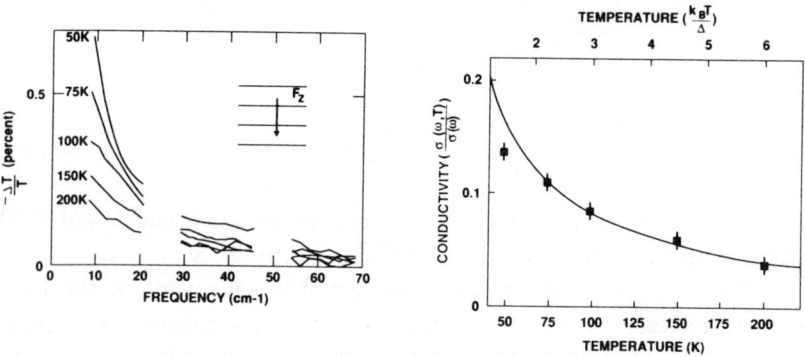

Fig. 2.6. Left: Drude conductivity of miniband transport in a narrow miniband SL. Right: Measured temperature dependence of the Drude conductivity (symbols) and calculated T-dependence (solid curve) using the relaxation time approximation (from Ref. 18).

only 3 meV wide. The concept of the miniband is indeed expected to break down when Δ becomes very small due to growth imperfections and other causes of departure from ideality. One first explanation could be the enhanced one-dimensional character of this experiment, because of the extreme quantum limit in a high magnetic field. However, such temperature variations were also independently evidenced by ordinary transport measurements in equivalent narrow miniband SLs[19] as discussed below. Secondly, the temperature dependence predicted by the relaxation time approximation implicitly neglects that of the scattering time τ. It is well known that many scattering processes such as phonon scattering exhibit appreciable temperature dependences. The good agreement with the relaxation time approximation therefore indicates a weak temperature dependence of τ. On account of the relatively low temperatures considered, it is likely that elastic scattering by interface roughness rather than phonon scattering was indeed the main temperature independent process in these experiments.

2.3.3. *Full Boltzmann equation*

The other three drawbacks quoted above of the Esaki-Tsu model require the solution of the full three-dimensional Boltzmann equation taking into account the main scattering mechanisms in the semiconductor structure considered. The equation is actually rendered two-dimensional by exploiting the cylindrical symmetry around the electric field direction, but is nonetheless demanding in terms of computation power. The Monte-Carlo technique, which is the most currently used in order to solve the Boltzmann equation at sufficient computing speed, has been applied to miniband transport by a few authors. Generally speaking, this method consists of defining a finite ensemble of particles subjected to the applied electric field in the semiconductor considered, which is defined by both its bandstructure and scattering processes. The ballistic acceleration and scattering events experienced by all particles of the ensemble are then individually calculated. For this purpose the occurrence times of the various collisions are chosen at random under the constraint of the known rates for each scattering process.

The result of such calculations for three GaAs/AlGaAs superlattices differing only by the Al composition in the barrier and a miniband dispersion approximated by the usual sinusoidal k-dependence[20] is shown in Fig. 2.7 footnotethere is an apparent reversal in the data points of Fig. 2.7 for the two extreme values of the miniband width, since the peak velocity is obviously an increasing function of this width.. Polar-optical, acoustical, and ionized impurity scattering processes *irrespective of the real superlattice structure* were taken into account, i.e., the scattering rates were those of an effective bulk material omitting SL effects like those related to the complicated features of the density of states.

The general appearance of the three $V(F)$ relations including the existence of NDV resembles that predicted by the crude Esaki-Tsu model or its improved version in the relaxation time approximation. However, it is instructive to compare the latter with the Monte-Carlo calculation quantitatively: for that purpose, the collision time τ was chosen by adjusting the low-field mobility in the two models. This procedure is justified by the pertinence of the relaxation time approximation *in the linear regime* when the collisions are elastic. The critical field predicted by the relaxation time approximation is then directly deduced from the relation $F_c = \hbar/(e\tau d)$. The agreement is acceptable, which means that the right order of magnitude of velocities and low-field mobilities is calculated by the relaxation time

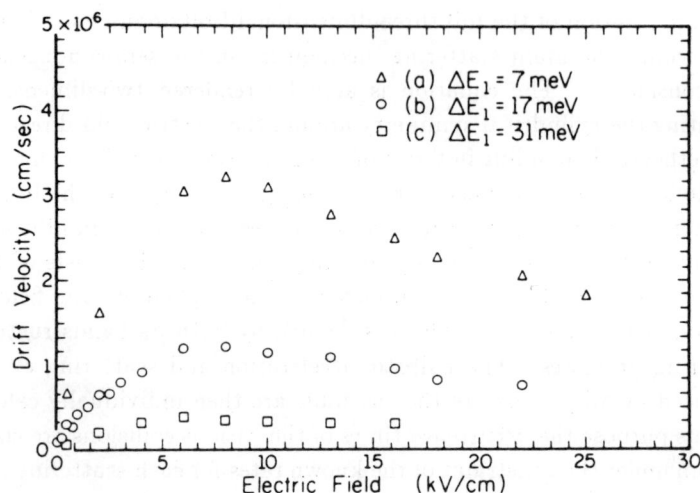

Fig. 2.7. Electron drift velocity vs electric field obtained by Monte-Carlo computations in three GaAs/AlGaAs superlattices with miniband widths from 7 to 31 meV (from Ref. 20).

approximation, but only through an ad-hoc adjustment of the *a-priori* unknown phenomenological time τ (cf. Tab. 2.2). This procedure is obviously unsatisfactory and a full solution of the Boltzmann equation is still needed for a critical analysis of the experimental data. The Monte-Carlo technique is powerful, but it requires a large set of particles in order to obtain significant results, in particular when subtle details are investigated which might be masked by statistical noise.

Table 2.2. Comparison between the main parameters of the $V(F)$ relation calculated either by the Monte-Carlo technique or through the relaxation time approximation for three miniband widths.

Δ (meV)	7	17	31
Monte-Carlo Model (from Ref. 20)			
μ (cm^2/(Vs))	150	600	1000
F_c (kV/cm)	6	7	8
V_p (10^{6}cm/s)	0.4	1.3	3.2
Relaxation time approximation			
I_1/I_0 $(\Delta/(2k_BT))$	0.067	0.162	0.287
τ (ps)	0.18	0.24	0.12
predicted F_c (kV/cm)	4.5	3	6
predicted V_p (10^{6}cm/s)	0.24	0.94	3.1

All these sophisticated models nevertheless demand a precise modelling of the collision phenomena. Three main mechanisms contribute to scattering in superlattices and may each limit SL conduction according to the temperature and sample characteristics. They are namely ionized impurity scattering, phonon scattering, and interface roughness (IFR) scattering. While the first two scattering mechanisms are not specific to superlattices, this is not the case for the third. This mechanism may indeed be expected to dominate scattering particularly in short period, high barrier SLs because of the large magnitude of the interface density/roughness potential.

The latter has been calculated by a few authors, under simplifying assumptions for the nature of the roughness. Chomette et al.[21] have considered one monolayer circular interface defects responsible for a local increased (decreased) well (barrier) thickness. The SL axis and in-plane low-field mobilities have been computed in the Born approximation for independent defects as a function of the defect radius (Fig. 2.8). It is important to note that in order to conserve a fixed total defect to SL volume ratio, the defect density had to be reduced for an increasing defect radius. This indeed appears physically sensible and agrees with common intuition of the existence of a roughness scale.

The plots reproduced in Fig. 2.8 show a minimum of the mobilities, i.e., *a maximum of the scattering* for a certain defect radius. This minimum could be explained by a transition between diffuse scattering (small defects) and specular scattering (large defects) which is predominant when the electron wavevector is a sufficiently good quantum number for k-selection rules to apply. The consequence of this remark is that it is preferable to have *either very poor interfaces or very good interfaces* in order to minimize scattering in the miniband. The first case is explained by the averaging effect of the electron wavefunction delocalization in the case of a finite wavevector.

One drawback of Chomettes's model was the unrealistic shape of the interface defects, while nature's well-known love for disorder is expected to favor irregularly shaped islands of various sizes at the interfaces. Dharssi and Butcher described this roughness by a Gaussian autocorrelation spectrum of the scattering matrix element between miniband Bloch functions parametrized by the coherence length of the IFR.[22] If one relates this length to Chomette's defect radius, a similar conclusion is indeed found with regard to the increased importance of scattering, when either of these parameters ranges from 1 to 100 nm (cf. Ref. 22). In any case, let it be phonon or IFR scattering, the mobilities are found to decay rapidly with the SL period as expected from the rapid variation of the miniband width with d (Fig. 2.9).

Dharssi and Butcher's model was extensively used by Palmier et al. in full Boltzmann calculations of non-linear miniband conduction.[23] We show for instance the dependence of the $V(F)$ relation on the well width a in several GaAs/AlAs superlattices keeping the miniband width constant. It shows *a dramatically increased scattering for narrower wells*, depending approximately quadratically on a^{-1} (Fig. 2.8). This behavior probably

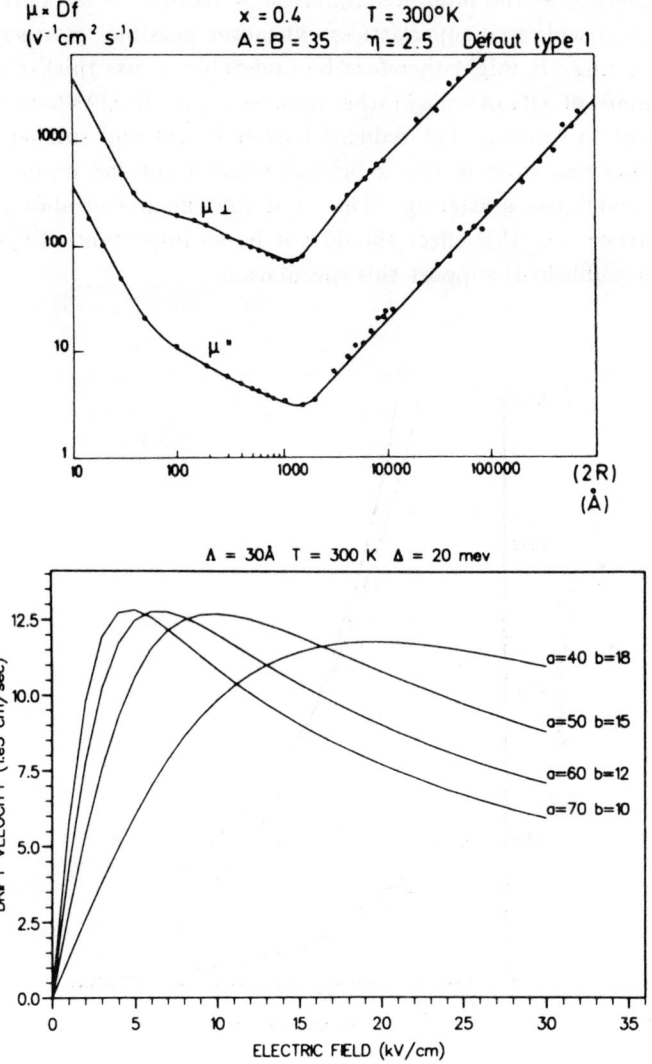

Fig. 2.8. Top: Longitudinal and transverse low-field mobility vs interface defect size in a superlattice calculated for cylindrical defects (from Ref. 21). Bottom: Well width dependence of the $V(F)$ relation in the case of strong interface roughness scattering calculated using the Boltzmann equation for a fixed miniband width (from Ref. 22).

results from both the high interface density and the enhanced electron probability density at the interface in narrow wells SLs. It strongly suggests the design of wide well superlattices, whenever possible, in order to minimize scattering. It might therefore be preferable to use thicker and lower barriers made of AlGaAs alloy rather than very narrow AlAs barriers, which are difficult to control. The reduced barrier height and smaller island to barrier thickness ratio is also a favorable factor for the minimization of interface roughness scattering. This is at the expense of alloy scattering in the barrier, but this effect should not be so important. Experimental results (unpublished) support this speculation.

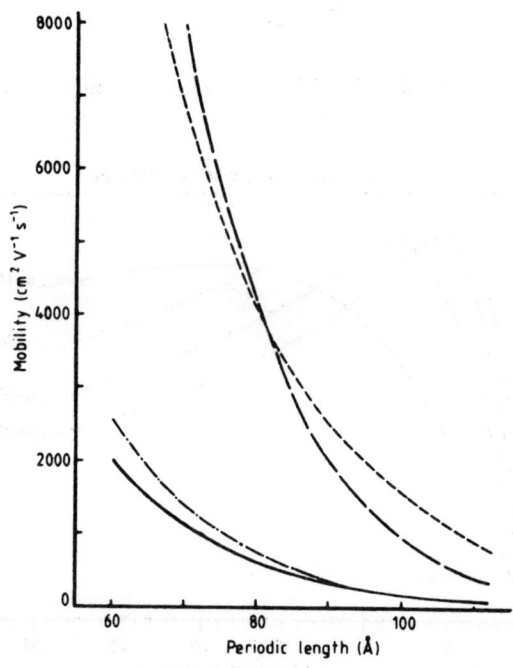

Fig. 2.9. Theoretical low-field electron mobility vs SL period in GaAs/AlGaAs superlattices plotted for various scattering mechanisms: LO phonon (long dashed line), IFR with roughness parameter 3.5 nm on one interface (short dashed line), IFR with roughness parameter 20 nm on the second interface (dash-dotted line), total IFR scattering (solid line).

2.4. Experimental techniques providing evidence of miniband transport

Since the initial proposal of artificial superlattices in 1970, many investigations have been devoted to the experimental demonstration of miniband transport, although the main unequivocal results were obtained as late as the end of the 1980's and thereafter. The main problem was naturally the extreme difficulty of growing well-controlled periodic stacks of heterostructure multilayers, precise on the monoatomic layer level. The impressive progress in molecular beam epitaxy techniques has made this possible, in particular in the GaAs/AlGaAs lattice-matched material system. In this section we briefly review some experiments, which allowed to access the properties of miniband conduction and determine some of the key quantities involved in miniband transport.

2.4.1. *Optical techniques*

Surprisingly enough, evidence of miniband superlattice conduction in the linear regime is rather scarce, possibly because of the absence of spectacular effects in this case as opposed to non-linear transport. Capasso et al. claimed miniband conduction to be the origin of a new quantum photoconductivity by effective mass filtering in forward-biased superlattice p-n junctions.[24] The basic idea was to take advantage of the large ratio of electron to heavy-hole effective masses in direct III-V semiconductors. Indeed one can expect to achieve a much faster electron transport than hole transport perpendicular to the layers, due to the sensitivity of the miniband width to the effective mass. In this case the rapid sweep-out of photoexcited electrons as opposed to the unusual slowness of holes was expected to be responsible for extremely high photoconduction gain, unfortunately at the expense of the device response time. True miniband transport is, however, not essential in this scheme, and the same authors found identical effects in a narrower electron miniband SL for which electron transport by hopping between disorder localized states was a more plausible mechanism.[25] Furthermore, it is now clear that there is a major contribution of hole capture due to particularly efficient hole traps in the high photoconduction gain experimentally measured, which dominates over slow hole transport and is responsible for quite low bandwidths and excitation-dependent gain.[26] For these reasons effective mass filtering is not a good solution to determine miniband transport in SLs quantitatively unless performing time-resolved experiments down to the picosecond time-scale.[27]

Optical techniques still have been employed in several ways in order to demonstrate miniband transport, basically for their inherent spectroscopic capability. Deveaud and co-workers have extensively investigated the photoluminescence of as grown superlattices in the aim of accessing perpendicular electron and hole transport by use of specially designed structures.[28,29] The seminal idea was to include in the SL one enlarged well which served as a very efficient intrinsic trap for photoexcited carriers or excitons, and thus could be used as an internal probe of electron (hole) miniband transport. In the simplest version of the technique, it is informative enough to measure the ratio of the enlarged well luminescence to the superlattice luminescence, which in a first approximation scales more or less with the ratio of the recombination and transport times along the SL growth axis.[30] However, the actual situation is rather intricate due to a number of phenomena and unknown parameters:

i) both electrons and holes are involved, which implies an important role of excitons in the lightly doped SLs used. It is therefore not *a-priori* evident that electron, hole or ambipolar transport is measured. This difficulty can be relaxed if a systematic study in samples of various doping levels is attempted.[31]

ii) not only the transport and the recombination times, but also the times of exciton formation and capture in the enlarged well are involved. This results in a dependence of the kinetic equations on a considerable number of parameters, among which only a few are *a-priori* well-known.

iii) the electric field in the structure is seldom reliably known because of the optical excitation and of the low doping level. Again variations in the sample design can help to reduce the uncertainty.

Time-domain techniques are powerful means for obtaining additional data in order to directly distinguish between the various times mentioned above. Particularly noticeable in this respect is the direct observation of carrier motion along the SL axis through spectrally resolved photoluminescence measurements. The 'stepwise graded-gap superlattice' samples were designed in such a way to produce a well-defined miniband in a variable gap structure so that the luminescence energy could be directly correlated with the exciton position in the SL. In this case the time-resolved photoluminescence spectrum could be simply related to ambipolar transport along the growth axis. Effective mobilities up to 1800 cm^2/(Vs) were deduced for

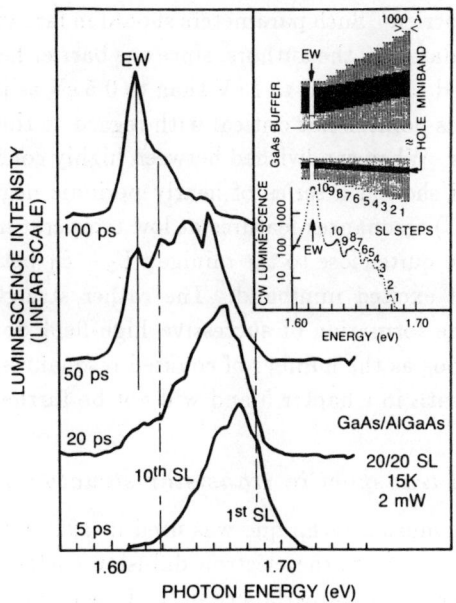

Fig. 2.10. Time-delayed photoluminescence spectra of a graded-gap superlattice demonstrating spectroscopically the fast sweepout of photo-carriers. Inset: Superlattice structure and CW luminescence showing as many peaks as the number of SL steps (from Ref. 28).

a GaAs/AlGaAs superlattice, indeed really indicative of miniband Bloch transport (Fig. 2.10).

2.4.2. *Miniband resonances in transport*

A few experimental investigations have been devoted to the observation and characterization of resonance effects in electrical transport due to the presence of one or several minibands. These effects have their origin in the succession of minibands and minigaps in the energy spectrum, which can be probed by sweeping the SL Fermi level E_F along them and observing enhanced electron quantum reflection, when E_F aligns with one minigap. The very first unambiguous demonstration of this was obtained by Esaki and Chang in 1974, probably as a by-product of their search for NDV.[32]

These pioneers used a Si-doped GaAs/AlAs SL (10^{17} e$^-$/cm^3) with 4.5 nm wells and 4.0 nm barriers. Such parameters should in fact yield narrower minibands than calculated by the authors, since the barrier height ΔE_C is now universally accepted to be closer to 1 eV than to 0.5 eV as it was assumed in 1974. However, this point is not critical with regard to the existence of the SL quantum states. When sandwiched between highly conducting contacts, the devices indeed showed a series of nearly periodic negative differential conductance (NDC) resonances features at low temperatures, separated by a voltage difference quite close to the minigap $E_2 - E_1$ between the ground state and the first excited miniband. The rather straightforward explanation involved the formation of successive high-field domains extending over as many periods as the number of counted resonances. This subject is extensively dealt with in Chapter 5 and will not be further discussed here.

2.4.3. *Miniband transport in transistor structures*

A somewhat unusual technique was used in 1986 by Palmier and co-workers in order to measure the electron diffusion coefficient along the SL growth axis by inserting the superlattice in the base of an n-p-n heterostructure bipolar transistor.[33] It is well-known that the common-emitter gain in such transistors made of direct III-V semiconductors is mainly limited by the base transport factor because of the near unity injection efficiency resulting from the presence of the emitter-base heterojunction. The transistor gain is therefore primarily determined by the recombination time at the emitter-base junction periphery or in the base itself, and by the electron diffusion coefficient in the direction from the emitter to the collector. When the recombination time of minority carriers is exactly known, a simple transistor gain measurement can thus easily yield the diffusion coefficient perpendicular to the transistor emitter-base heterojunction, or equivalently yield the mobility in the base through Einstein's relation. If the base itself is a superlattice, this technique can thus provide direct evidence of miniband conduction, if large perpendicular mobilities only compatible with scattering limited transport between delocalized Bloch states are measured. Generally, recombination times are only guessed approximately so that a full set of computer simulations of the transistor characteristics was needed in the cited experiments[33] in order to deduce electron and hole mobilities together with recombination times. Within reasonable bounds of the latter, a high electron perpendicular mobility of about 1000 cm^2/(Vs) in the SL was found in one 4.5 nm/2.0 nm $GaAs/Al_{0.27}Ga_{0.73}As$ SL, while a much lower mobility below 10 $cm^2/(Vs)$ was determined in 4.0 nm wide barrier SL (Fig. 2.11). This very dramatic difference in otherwise identical devices could therefore be interpreted as a strong indication of the existence of a transition from Bloch miniband to hopping transport between disorder localized states, when the barrier width increases, i.e., when the calculated miniband width drops below a certain threshold value. Since that time, progress in the microscopic quality of superlattices seems to have pushed this threshold lower, as will be discussed below.

Following another approach, England et al. subsequently investigated superlattice transport in *unipolar* hot electron transistors, i.e., based on pure electron conduction in all n-type structures.[34] Such devices have been demonstrated in bulk semiconductors like GaAs since the beginning of the last decade and operate in a manner resembling somewhat bipolar transistor action. An *emitter* has the role of injecting electrons towards a *base* through a wider bandgap barrier. Since the electrons are injected with an energy appreciably greater than k_BT, they can move nearly ballistically in the base and eventually reach a *collector*, if their residual energy allows them to surmount a second barrier. The usual recombination of electrons with holes of the base in a conventional transistor is here replaced by the very fast relaxation of these 'hot' electrons towards the bottom of the conduction band, where they are inhibited to reach the collector. In order for the device to exhibit gain, it is necessary that the hot electron mean free path was longer than the base thickness. Thin and, consequently, appreciably doped bases are therefore required.

Because the injection energy can be modulated by the applied bias on the emitter-base junction, it is in principle possible to spectroscopically investigate transport in the ground *and also in the higher minibands* of a superlattice by embedding the SL in the base of a unipolar hot electron transistor and looking at the bias dependence of the transistor conductance.[34] Such experiments conducted on GaAs/AlGaAs SLs clearly showed a series of resonances in the transistor conductance, which corresponded to injecton into successive minibands of the superlattice. At the origin of these resonances was the quantum reflection experienced by the electrons whenever their energy fell within one minigap of the SL base, thereby reducing the injection efficiency. From the experimental determination of the gain, the mean free path in each miniband can thus be estimated. A rather low gain was actually measured and short mean free paths of about 3 SL periods were deduced due to particularly short scattering times of about 20-50 fs. Such low values make the very existence of a miniband somewhat questionable, since rigorously it requires the wavefunctions to be coherent over length

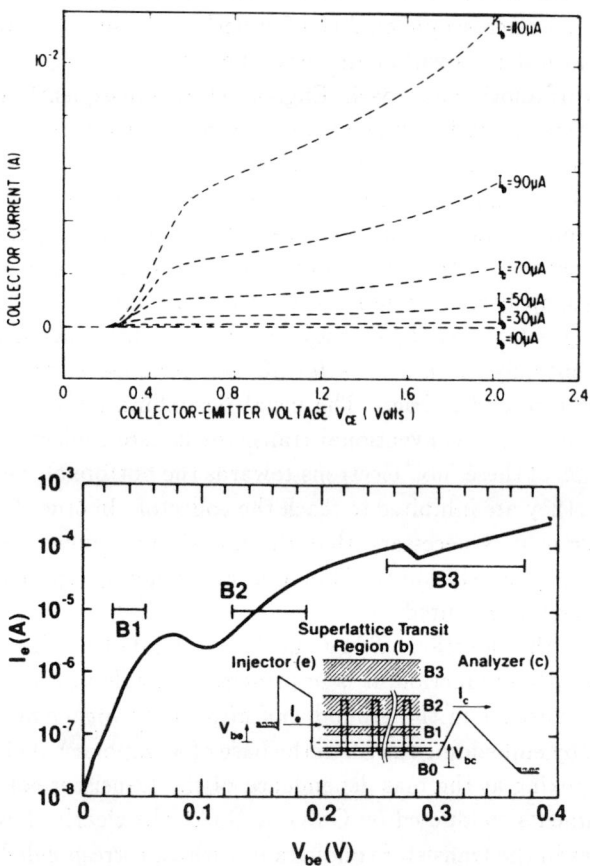

Fig. 2.11. Top: Common-emitter characteristic of a superlattice bipolar transistor. The gain of 400 indicates a diffusive mobility in the SL base of about 1000 cm^2/(Vs) (from Ref. 33). Bottom: Tunneling injection into higher order minibands of a superlattice base unipolar hot electron transistor (from Ref. 34).

scales much greater than the SL period. It is certainly rather more appropriate in the present case to view these effects as resonances induced by the presence of (lifetime broadened) quantum states, defined on a very small number of wells.

2.4.4. *Experimental demonstration of miniband NDV through d.c. current-voltage measurements in unipolar structures*

2.4.4.1. *Undoped superlattices*

The experimental demonstration of NDV in a biased superlattice is not as easy as it could seem at first sight. The reason for this connects with the interplay between the non-linear $V(F)$ relation and the overall charge distribution through the transport equations. This interplay is at the origin of the dramatic electric field inhomogeneities throughout the sample. For this reason it is illusory to directly extract $V(F)$ relations such as predicted by one of the various NDV models discussed above, from current-voltage I-V data. In the following we therefore detail a number of experimental attempts to measure $V(F)$ relations, which have been conducted in undoped GaAs/AlAs superlattices sandwiched between GaAs n^+-contacting layers.[35]. Such n^+-p-n^+ structures are not prone to easy interpretations of I-V data, so that the full self-consistent solution of the coupled transport and the Poisson equation was estimated to be an indispensable way to obtain a reliable characterization of miniband transport. In order to reach this goal, it is indeed necessary to describe the entire device realistically, since the boundary conditions at the superlattice end have to be determined precisely. Furthermore, it is essential to dismiss any spurious heterojunction effect between the SL and the access layers, which would otherwise mask or, at least, affect SL transport measurements through a bad electrical access to the SL. In order to remove these 'quasi-heterojunctions', it was found to be practical to insert between the SL and the GaAs contacts an intermediate layer made of a pseudo-alloy chirped short period superlattice. The parameters of the latter were regularly graded in such a way as to have a continuous variation of the conduction miniband minimum from the SL up to GaAs. This electron injector structure was modeled as an average graded composition $Al_{1-x}Ga_xAs$ bulk material. However, an analytical expression for the superlattice $V(F)$ relation still has to be assumed in order to be able to compute I-V characteristics. Early investigations have made extensive use of the expression

$$V = \frac{\mu F}{1 + (F/F_c)^\eta} , \qquad (2.15)$$

which has the quality of covering all important cases with regard to SL transport: ohmicity directly occurs for $\eta = 0$, a saturating $V(F)$ law is found if $\eta = 1$, and finally NDV is present, if $\eta > 1$. The Esaki-Tsu model is also described by this law by taking $\eta = 2$. In the case of NDV beyond the critical field F_c, we have

$$F_c = F_0 \left(\frac{1}{\eta - 1}\right)^{1/(\eta-1)} . \qquad (2.16)$$

In undoped samples the content N_a of residual acceptor impurities is *a-priori* unknown so that a full adjustment between experimental and computed I-V data requires the fitting of N_a, μ, F_0, and η. Although it could seem illusory to attempt deducing significant values of all four parameters, their influence on the I-V characteristics is quite specific in each case: while the I-V is *linearly* modified by μ, it is actually a strongly non-linear function of N_a, especially under low bias. This is because N_a determines the amount of band-bending in the SL due to ionized acceptors. The acceptors are responsible for a **space-charge energy barrier for electrons** and have to be compensated by an applied voltage in order to have any chance of passing an appreciable current through the device. Stated in simple words, the n$^+$-p-n$^+$ SL structure behaves as a bipolar transistor with a too lightly doped and uncontacted base. The space-charge barrier requires for its removal a sufficiently large applied bias in order to obtain the so-called 'punch-through' regime, in which F_0 and η are finally the most important parameters.

Adjusting all four parameters consequently needs an important effort, but it is rewarded by the physical information contained in the very shape of the $V(F)$ relation. We show in Fig. 2.12 the dependence of the computed I-V on the $V(F)$ relation for one sample in comparison with experimental data. A simple saturation law was first assumed (curve d corresponding to $\eta = 1$), as well as an Esaki-Tsu law ($\eta = 2$) for three different values of the critical field $F_0 = F_c$. It is apparent that a satisfactory fit could only be obtained for a given value of F_c, namely $F_c = 16.5$ kV/cm in this 13/7 monolayer GaAs/AlAs superlattice. Particularly important is *the sublinear*

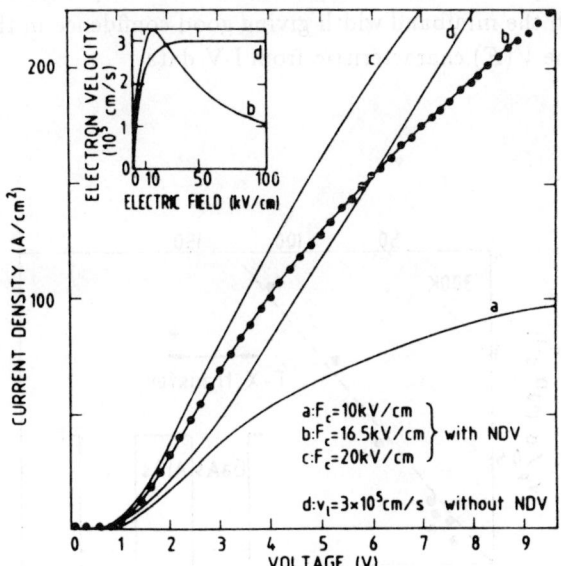

Fig. 2.12. Experimental (dots) and simulated (full lines) I-V characteristic of a GaAs/AlAs superlattice. An excellent agreement could be obtained only in the case of NDV among various $V(F)$ relations such as shown in the inset (from Ref. 36).

I-V characteristic, which is a feature betraying the existence of NDV. In the cited example, a low-field mobility of 40 cm^2/(Vs) was also determined together with a residual acceptor content of $N_a = 1.4 \times 10^{15}$ cm^{-3}. The method was also validated by direct comparison between experimental and computed I-V characteristics of a GaAs sample, for which the $V(F)$ law was known from the literature.

As an example of the experimental generalization of this early result, a plot of the peak velocity divided by the SL period and of the voltage drop per period at the critical field are shown in Fig. 2.13. These quantities rather than more naturally V_p and F_c were chosen in order to compare them with the Esaki-Tsu model described above which yields $V_p/d = \Delta/(4\hbar)$ and $edF_c = \hbar/\tau$. The first plot, although linear, is actually about three times lower than predicted by this crude model, while the second seems to indicate a sample-dependent scattering time τ. The marked limitations of the model are here exemplified, but the emphasis is on the very regular variation of the data with the miniband width giving good confidence in the extraction method of the $V(F)$ characteristic from I-V data.

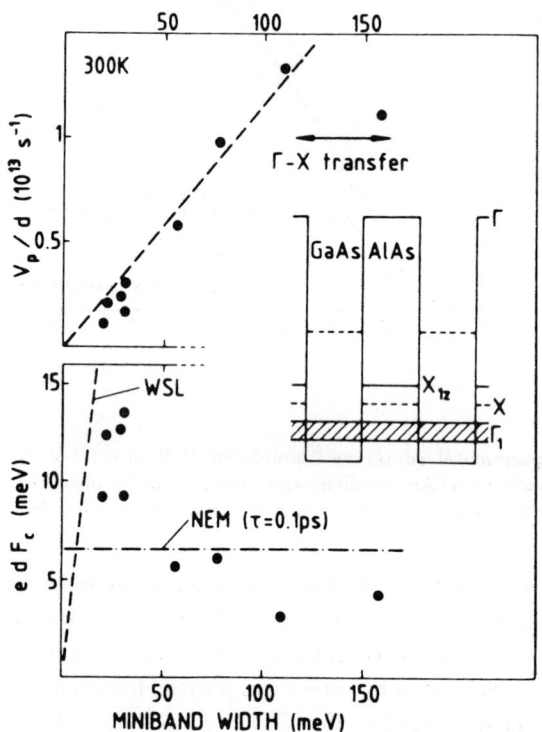

Fig. 2.13. Top: Experimental peak velocity vs miniband width in a series of GaAs/AlAs superlattices. Bottom: Product of the critical field with the SL period in the same samples compared to the Esaki-Tsu model for a fixed collision time (from Ref. 35).

2.4.4.2. *Doped superlattices*

The situation in doped superlattices is qualitatively different from undoped ones because of the large amount of mobile charge susceptible of being involved in the solution of the coupled charge-transport equations. This problem relates to the more general question of the existence and stability of the solution in an electrically contacted medium exhibiting negative differential mobility, which has been treated both experimentally and theoretically in the past with application to the Gunn effect. In this

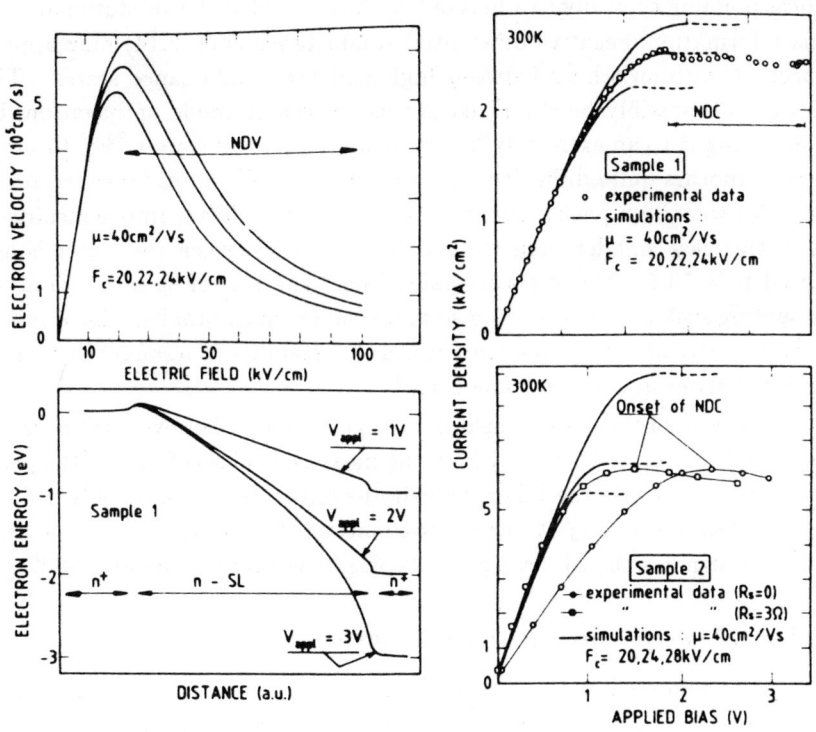

Fig. 2.14. Left: Typical $V(F)$ relations and computed conduction band diagrams under bias in lightly doped n^+-n-n^+ SL structures. Right: Experimental room temperature I-V data of two n^+-n-n^+ GaAs/AlAs superlattices together with simulated characteristics (from Ref. 38).

context Guéret has demonstrated analytically the possibility of both *absolute* and *convective* instabilities depending, respectively, on the doping level being larger or smaller than a certain critical value determined by the characteristic dependence of the electron velocity on the field in the NDV regime.[37] In the second case ('high' doping), any local disturbance will be amplified in such a way as to develop a non-uniform electric field domain located on the anode side of the medium, at the ohmic contact boundary. We show in Fig. 2.14 the computed diagrams under several applied biases of the bottom of the conduction miniband in one SL exhibiting NDV.[38] They indeed clearly show the development of a localized high field domain on the anode side, when the average electric field over the SL exceeds F_c corresponding typically to a bias of 2 V. Associated with this stationary domain formation, negative differential conductance (Fig. 2.14) may appear together with switching between high and low conductance states. The device can possibly be stabilized against travelling mode propagation, by connecting it to an extremely low impedance external circuit.[39,40] In practice it appears quite difficult to obtain such a stabilization for superlattice samples with high peak current densities J_p. The device impedance itself is in this case smaller than the source impedance, when the latter is adapted to a 50 Ω coaxial transmission line. The role of spurious cable or mounting system parasitics is also appreciable. Such problems have raised a controversy about the so-called intrinsic bistability of resonant tunneling double-barrier devices for some time.[41]

In spite of the electric field inhomogeneity, it is however still possible to directly estimate V_p and F_c from the measured values of J_p and the peak voltage U_P, because the inhomogeneity begins to be dramatic only *beyond* F_c. The simple *rule of thumb* approximations $J_p \approx neV_p$ and $F_c \approx U_p/L$, where L denotes the SL length, are typically accurate to about 10-20%.

2.4.5. *Magnetotransport measurements*

The application of a magnetic field in addition to an electric field is a commonly adopted way of gaining insight into the physical processes responsible for electron transport. In the case of superlattices, the highly anisotropic shape of the Brillouin zone is expected to greatly affect the semiclassical dynamics of electrons subjected to both fields, and to lead to unusual magnetotransport effects. Suris and Shchamkhalova have predicted negative magnetoresistance along the SL layers under low electric fields,[42] while Mitra and Ghatak have calculated the departure from Einstein's relation in degenerately doped superlattices under large crossed magnetic and electric fields.[43] When the magnetic field is low the semiclassical approximation still holds, according to which Landau quantization of the electron states can be safely neglected. This applies when the collisional broadening energy $\hbar/\tau$ is greater than the inter Landau level spacing. The electron trajectory $\vec{k}(t)$ in wavevector space is governed by the semiclassical equation

$$\hbar \frac{d\vec{k}}{dt} = e(\vec{F} + \vec{v} \times \vec{B}) \,. \qquad (2.17)$$

2.4.5.1. *Magnetotransport in crossed fields*

In the following we focus on magnetotransport in the crossed field configuration (F parallel to the SL axis and B parallel to the layers). It is easy to show that the trajectories for electrons in a miniband are two-dimensional and can be separated in two classes, open and closed orbits. The transition between both of them actually occurs by changing either F or B, and is achieved when $B = F\sqrt{m_{x,y}^*/(2\Delta)}$, where $m_{x,y}^*$ denotes the effective mass in the layer plane. In a first approximation, the B-dependent $V(F)$ relation can be obtained by extending the Esaki-Tsu model to two dimensions using

$$< V > = \frac{\int_0^\infty V_g(\vec{k}(t)) \, exp(-t/\tau) \, dt}{\int_0^\infty exp(-t/\tau) \, dt} \,. \qquad (2.18)$$

The numerically computed $V(F)$ relations are shown in Fig. 2.15 and Fig. 2.16 in comparison with experimental I-V characteristics for one GaAs/AlAs SL with a miniband width of about 20 meV. There is only a rough qualitative agreement between both plots, which is not surprising in account of the simplicity of the model and of the neglect of the electric field inhomogeneity implicit in such a comparison.

It is interesting that a plot of the velocity or current vs B shows a positive magnetoresistance under low electric fields, *while a negative magnetoresistance is first observed when the electric field is large enough, i.e., in the NDV regime.* This magnetoresistance peak interestingly shifts linearly with the applied electric field, which is strongly suggestive of the role played by the above transition between open and closed orbits. Aristone et al. have confirmed this assignment by investigating a series of

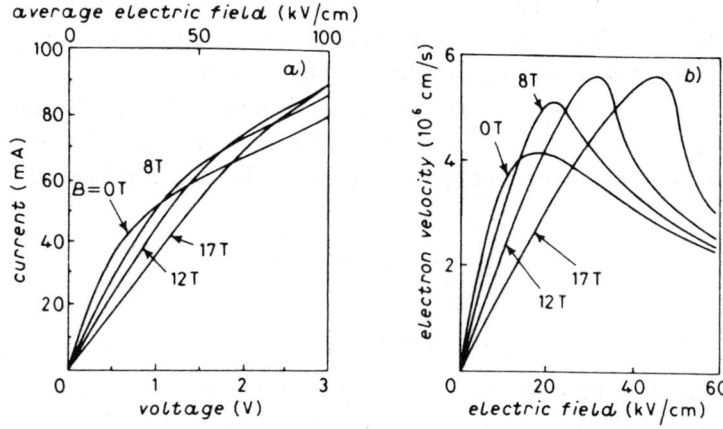

Fig. 2.15. a) Experimental I-V characteristic of a lightly doped GaAs/AlAs superlattice under various in-plane magnetic fields. b) $V(F)$ relation calculated using a modified Esaki-Tsu model for the same magnetic fields. Although electric field inhomogeneities prevent direct comparison, the same general trends are observed (from Ref. 46).

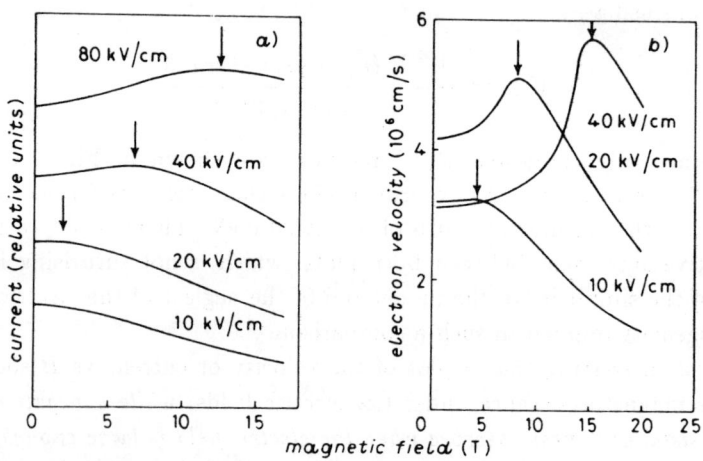

Fig. 2.16. a) Experimental current vs magnetic field corresponding to the data of Fig. 2.15. b) Calculated magnetic field dependence of the electron velocity (from Ref. 46).

samples with different miniband widths in quite good agreement with the $1/\sqrt{\Delta}$-dependence cited above.[44] Actually a much better agreement between experiment and theory was obtained by Palmier et al. by solving the Boltzmann equation accounting for the main scattering mechanisms together with the Poisson equation in a self-consistent manner in order to fully take into account the electric field inhomogeneity (Fig. 2.17). Finally, it is worth mentioning the work by Kop'ev et al.[45] who observed *hole* localization by a magnetic field applied in the layer plane of GaAs/AlGaAs superlattices. They used optical probing of miniband transport with an enlarged well as described in section 2.5.1, and found a drastic increase of the SL luminescence in a magnetic field in the case of narrow hole minibands. This effect betrayed hole localization which is understood as the result of the breakdown of the miniband under fields such that the cyclotron energy exceeded the hole miniband width. Electrons were not directly involved, since their miniband width would have required much larger magnetic fields.

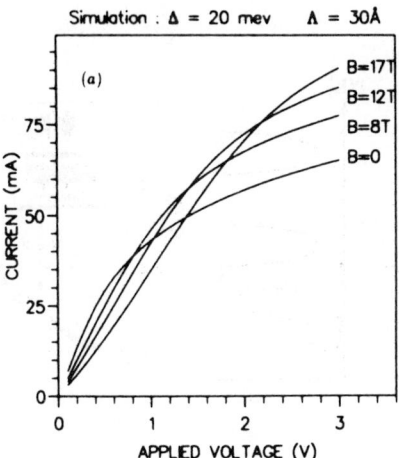

Fig. 2.17. Computed I-V characteristics of the superlattice of Fig. 2.15 using both, the Boltzmann and the Poisson equation (from Ref. 48).

2.4.5.2. *Magnetotransport in parallel fields*

The application of a large magnetic field along the SL growth axis provides another way of studying quantum miniband conduction effects through Landau quantization of the electron orbits in the layer plane. Because of the separation between longitudinal and transverse motion in this field configuration, the energy dispersion reads

$$E(\vec{k}) = E_z(k_z) + (n + \frac{1}{2}) \hbar \omega_c , \qquad (2.19)$$

where $\hbar\omega_c$ denotes the cyclotron energy. In the extreme quantum limit this energy appreciably exceeds the miniband width, which results in true one-dimensional conduction. Noguchi et al.[47] have considered the case of degeneracy $N\hbar\omega_c = \hbar\omega_{LO}$ between the cyclotron energy and one or several longitudinal optical phonon energies $\hbar\omega_{LO}$ to induce longitudinal magnetophonon resonances in transport (Fig. 2.18). Electrons accommodated in the lowest Landau level can be excited to higher ones by optical phonon

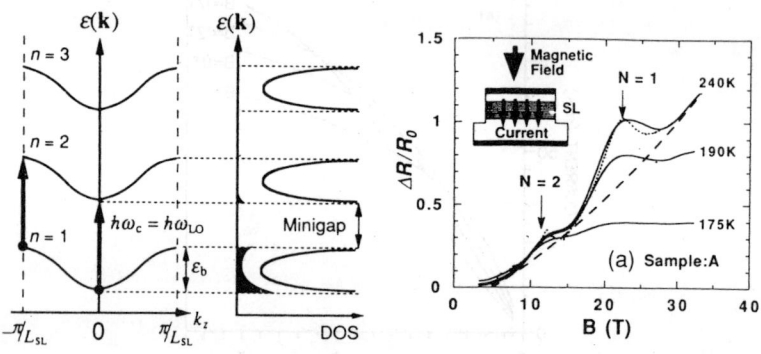

Fig. 2.18. Left: Dispersion relation and density of states of a superlattice in a large magnetic field. Right: Longitudinal magnetoresistance of a superlattice at different temperatures showing prominent magnetophonon resonances (from Ref. 48).

scattering resulting in many resonances in the magnetic field dependence of the device resistance (Fig. 2.18). This effect is magnified in superlattices, because the phonon scattering rate is sensitive to the density of states at the initial and final states, both exhibiting a singularity when the resonance condition is achieved. For this reason narrow miniband SLs are particularly favorable, since the magnitude of the singularity is enhanced. Experimental data indeed show a clear miniband width dependence of the amplitude and the shape of the resonances in the expected form.[47] This effect may usefully help to characterize phonon scattering at the temperatures of interest and to investigate the origin of the involved phonon modes. The spectroscopic capability of the technique through the magnetic field peak position may indeed be directly related to the phonon energy and eventually to its spatial dependence by correlating it with the microscopic design of the sample.[49]

2.4.6. *A.C. electrical characterization of miniband transport*

2.4.6.1. *Origin of the a.c. response*

Although the previously discussed static measurements of the SL conduction show direct evidence of NDC and allow a relatively precise estimation of the $V(F)$ relation, they rely, however, on heavy self-consistent computations of the transport and charge equations in order to obtain a good accuracy. At the same time they suffer from geometrical uncertainties of the measured device. In analogy with the normal operation of transferred electron devices, we can take advantage of the NDV/NDC in order to perform experiments *in the frequency domain*. The frequency response must indeed exhibit features related to propagative modes of a non-equilibrium perturbation of charge (electric fields) in a medium exhibiting NDV associated with an instability of convective nature.

The analytic calculation of the small signal, harmonic response to an applied microwave excitation in an n^+-n-n^+ structure with NDV has been carried out by Hakki in the case of the Gunn effect based on the transfer between a high and a low mobility valley.[50] The conclusions, which are anyway at least qualitatively relevant for NDV in superlattices, stress the existence of resonances in both the in-phase and out-of-phase response at those frequencies, which are integer multiples of the typical inverse transit time $< V > /L$, where $< V >$ denotes the average electron velocity. These resonances can be understood in terms of a delay-line effect for the transit of an extra injected packet of charge from the cathode to the anode through the SL of length L. The presence of NDV is required in order for the disturbance to equilibrium to be amplified, due to a negative dielectric relaxation frequency instead of being attenuated as it is normally the case for positive differential velocity.

However, the doping of the active layer is not a necessary ingredient for the successful observation of such resonances. Numerical simulations of the computed microwave response on undoped (p^-) superlattices sandwiched between n^+-contact layers have been performed.[51] Although the conductance remains positive in all cases, they indeed show very well defined resonances at the fundamental frequency and higher harmonics of the inverse average transit time $< V > /L$. Increasing the applied bias has the effect of decreasing $< V >$ and therefore of uniformly shifting all resonances to low frequencies, which is a direct manifestation of NDV.

2.4.6.2. *Microwave probing*

The conductance and capacitance of several p^- and n^- superlattice structures have been measured under microwave excitation by connecting them to a broadband network analyser. The results are in excellent agreement with the expected behavior discussed in the previous section (Figs. 2.19 and 2.20). It is quite apparent from the data that

i) the sublinearity of the I-V characteristic becomes more and more pronounced as the doping level is increased, and eventually results in true NDC.

ii) at the same time the microwave resonances sharpen, while their relative amplitude increases.

iii) the ample shift to lower frequencies of the resonance peak positions strongly confirms the superlattice NDV.

iv) the shift of the resonance frequency towards larger values, which scales *inversely with the SL thickness*, is a direct proof that these resonances and the associated NDV are a bulk superlattice effect, i.e. are not quantum resonances due to some localized defect of an imperfect structure. Indeed this experimental observation demonstrates the direct connection between the resonance frequency and the inverse transit time $< V > /L$ and therefore the involvement of the superlattice as a whole.[52]

It is clear that microwave investigations are more powerful with regard to the SL conduction than simple I-V measurements, because of the wealth

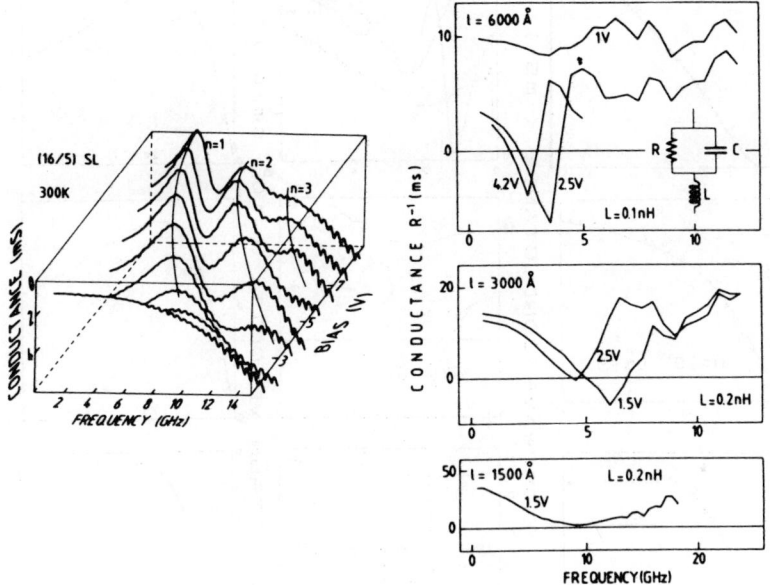

Fig. 2.19. Left: Bias dependence of the microwave conductance spectrum in an undoped SL (from Ref. 35). Right: Microwave conductance of three samples with identical SL structure showing the scaling between the resonance peak frequency and the inverse SL thickness l (from Ref. 52).

of information contained in the spectra. In particular, we can expect to accurately extract the electron velocity dependence on the electric field in the NDV regime from the resonance amplitude and shift with applied voltage, which can only be approximately guessed from the sole I-V. This is the subject of future investigations in order to fully characterize NDV miniband transport as a function of the superlattice parameters.

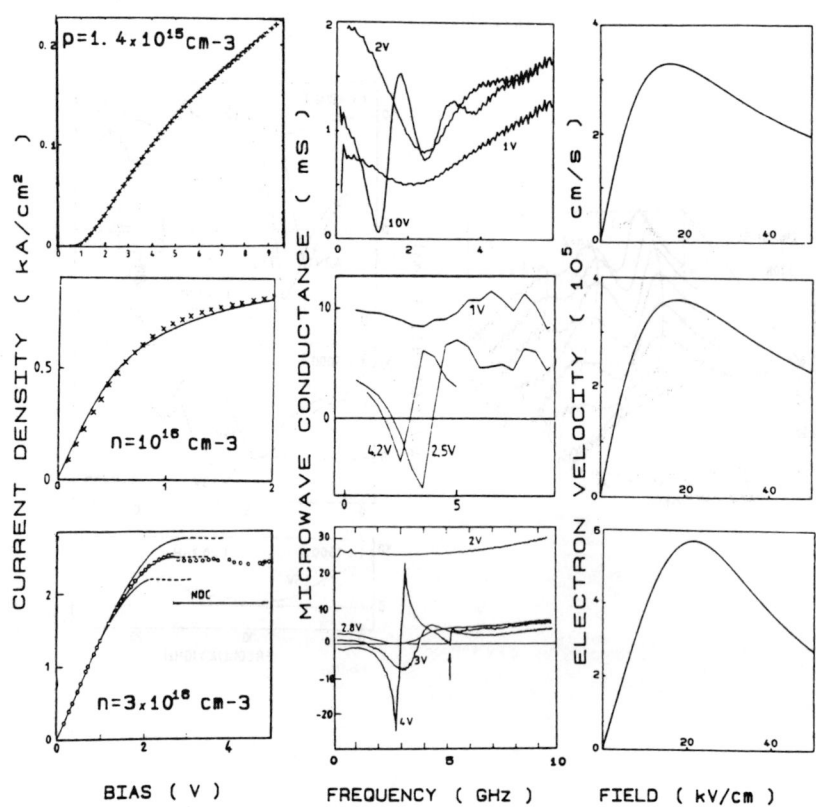

Fig. 2.20. I-V characteristic at 300 K, microwave conductance spectrum, and $V(F)$ relation for a series of GaAs/AlAs superlattices with increasing doping level (from Ref. 53).

2.4.6.3. *Time-of-flight probing*

A variation of the microwave technique for the direct observation of miniband transport can be carried out *in the time domain* through time-of-flight experiments. The latter consist of the sample illumination by a short light pulse followed by the time-resolved monitoring of the current response, which can be related to the drift of the carriers in the applied electric field. Since the first Shockley-Haynes experiment, this technique has been applied for many years in low mobility materials or in long samples. It can also be used in thin SL structures characterized by fast miniband transport under the conditions of both, an ultra-short pulsed photoexcitation and ultra-broadband detection of the device electrical response. This requirement imposes a double effort in the instrumentation and in the technological processing of the samples, which should minimize the parasitics susceptible to affect the overall response. Compared to microwave measurements, the time-of-flight technique has the drawback of combining electron and hole transport effects. However, this drawback turns out to be an advantage, if the velocity difference of electrons and holes and an accurate modelling allows for the reliable separation of both velocity components, leading to a simultaneous determination in a single sample and set of experiments.

In Fig. 2.21 the time-resolved photocurrent response following a picosecond light pulse incident on a 22 meV wide miniband GaAs/AlAs SL obtained by Minot et al.[27] is shown. The main peak is ascribed to the *primary* photocurrent, i.e., current resulting from the initial photocreated electron-hole pair. It is followed by a *secondary* transient due to the subsequent extra electron injection by the cathode lasting as long as the total recombination time in the SL. This is the well-known photoconductor effect responsible for a larger than one internal gain, expressed as the integrated photocurrent (in elementary charge unit) per absorbed photon. A computed photocurrent transient favorably compares with the data after adequate adjustment of the various parameters. The slow decaying tail is caused by recombination and hole sweepout characterized by typical times of a few nanoseconds. Very much like microwave experiments, the presence of the primary peak is by itself a signature of NDV as it is absent from simulated responses deprived of NDV. Furthermore, *the peak markedly shifts to longer time delays as the applied voltage is increased*, which is another manifestation of NDV. The same kind of effects were observed in time-resolved pump-and-probe experiments in superlattice p-i-n junctions.[54]

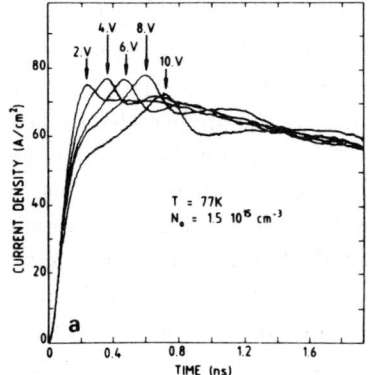 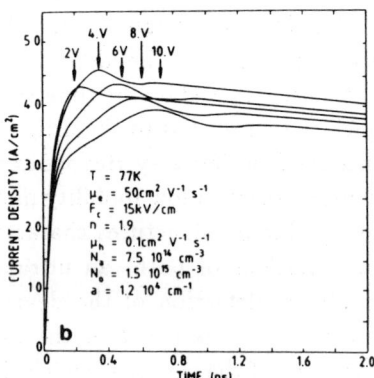

Fig. 2.21. Experimental (left) and simulated (right) time-resolved photocurrent in a superlattice showing the shift of the main peak to longer delays as the bias is increased (from Ref. 27).

2.5. Advanced analysis of miniband conduction

2.5.1. *Transport in wide minibands*

The Esaki-Tsu model and its slightly improved version in the relaxation time approximation are basically one-dimensional models, because there is a perfect decoupling between the longitudinal and transverse motion with respect to the direction of the applied field. Real superlattices are three-dimensional, because

i) the electron energy cannot be separated in longitudinal and transverse components as soon as $\vec{k} \neq \vec{0}$

$$E(\vec{k}) \; \neq \; E_{\parallel}(k_{\parallel}) \; + \; E_{\perp}(k_{\perp}) \,. \tag{2.20}$$

ii) the collision term in the Boltzmann equation explicitly couples different electron Bloch states irrespectively of their direction relatively to the electric field.

In the present section we focus on point ii), which is responsible for a transfer of energy from the longitudinal to the transverse motion. As a consequence 'heating' in the layer plane can be expected for minibands wide enough to allow an appreciable electron acceleration by the field. It is clear that the magnitude of heating is of paramount importance for the true shape of the $V(F)$ relation, in particular for the peak velocity and the value of the negative differential mobility in the NDV regime. This can be easily understood by the simple consideration of the $V(F)$ relation in the relaxation time approximation incorporating the factor $I_1(x)/I_0(x)$, where $x = \Delta/(2k_BT)$, which decreases rapidly at elevated (electron) temperatures.[55] This effect has been investigated theoretically by several authors. Levinson and Yasevichyute claimed that the increased dissipation at high field should cause a reduction of the velocity with F faster than $1/F$,[56] a conclusion which is not supported by Boltzmann computations.[23,57,58] It was also somewhat contradicted in analytical calculations by Suris and Shchamkhalova in phonon limited transport[17] indicating a maximum electron temperature $T_E = T_0[1 + (\Delta/\hbar\omega_{LO})^2]$ in the high field limit, where T_0 denotes the lattice temperature and $\hbar\omega_{LO}$ the optical phonon energy. This result can be understood from an energy balance condition between the rate of energy transfer to the transverse movement determined by the momentum relaxation rate and the rate of energy transfer to the lattice by phonon emission. The lower the phonon energy, the lower is proportionally the energy relaxation rate. Heating is also quadratically favored in wide miniband SLs as one intuitively expects. We show in Fig. 2.22 the comparison between the experimental data of Fig. 2.13 and calculations based on the balance equation by Lei et al.[57] using the pertinent SL parameters and one fitting parameter: the ionized impurity density per SL period. The agreement is impressive for both, the peak velocity and the critical field. However, the balance equation theory relies on strong electron-electron interactions in order to justify the existence of a thermalized distribution and an electron temperature. Since the investigated samples were undoped, the impurity scattering involved in the calculation should be regarded as an ad-hoc ingredient of the model. It is also surprising that the electron temperature calculated by these authors appears to increase without limits for large fields, since in the NDV regime well beyond F_c the electrons explore the whole Brillouin minizone and the average kinetic energy is limited to half the miniband width Δ. Full Boltzmann calculations, however, predict a saturation of T_E.[59] This point therefore needs to be elucidated.

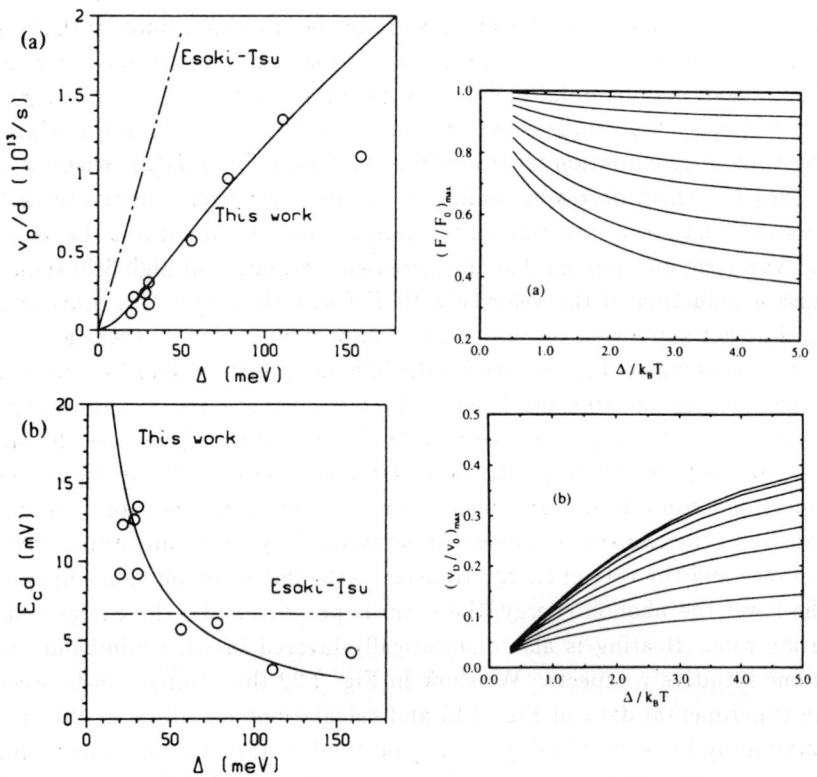

Fig. 2.22. Left: Comparison between calculations of superlattice $V(F)$ parameters obtained through the balance equation and the experimental data of Fig. 2.13 (from Ref. 56). Right: Critical field and drift velocity vs miniband width obtained using the Boltzmann equation with ratios of elastic to total collision rate increasing from top to bottom curves (from Ref. 57).

The regular decrease of F_c with Δ is another unexpected feature from the one-dimensional miniband conduction models. Gerhardts[58] has shown that it is a direct consequence of the importance of _elastic_ vs _inelastic_ scattering. Increasing the first with respect to the second favors lateral electron heating. Since the latter takes place for large electric fields, the velocity is more reduced for large than for small fields in comparison with the Esaki-Tsu model. In other words F_c *is effectively shifted to lower fields*, in particular for wide minibands (Fig. 2.22). This conclusion is illuminating for the understanding of the success met by Lei et al. in fitting experimental and calculated miniband conduction data in spite of the erroneous impurity scattering: the elastic character of the latter is indeed the important part. Any other elastic process would probably provide comparable results. Actually interface roughness scattering in the GaAs/AlAs small period SLs concerned is a much more realistic origin for this elastic contribution as discussed in section 2.4.3. Again it seems advantageous to minimize this kind of scattering, since it reduces the peak velocity still further than the low-field mobility (Fig. 2.22).

2.5.2. *Transport in narrow minibands*

If we except 'perpendicular' transport in multiple quantum wells, which cannot be qualified of superlattices, the conduction processes in narrow minibands have been investigated only a little. The experiments of Brozak et al. have been cited above, although they were conducted in the absence of any applied electric field except for that of infrared light. The agreement, which was observed with the relaxation time approximation as far as the temperature dependence was concerned *-including the miniband width dependence-*, is undoubtedly an important feature for the validity of this interpretation.

More convincingly, Grahn et al. performed time-of-flight experiments of the $V(F)$ relation in a series of narrow miniband SLs with widths between 0.5 and 1.5 meV.[60] Although the measurements were qualitative rather than fully quantitative because of the approximate method of data extraction, the experimental features were sufficiently strong to be significant (Fig. 2.23). They also observed clear negative differential velocity in the low temperature range up to at least 130 K. Again the decrease of the mobility with T in agreement with the relaxation time approximation was noted.

These results were extended and confirmed by a series of measurements performed on lightly doped n-type GaAs/AlAs superlattices of miniband widths between 4 and 16 meV differing only in the well/barrier thickness.[19] In all cases, NDV as well as negative differential conductance was found at room temperature and below (Fig. 2.24). This behavior also included well-defined NDC microwave resonances at harmonic frequencies

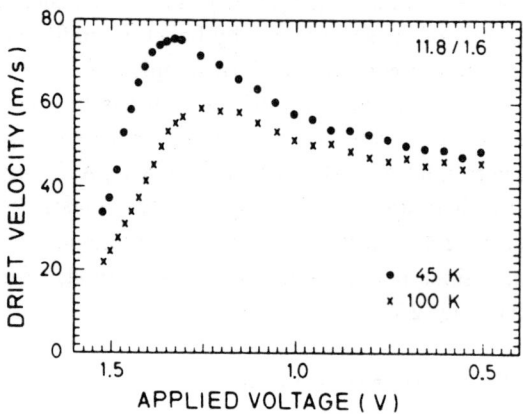

Fig. 2.23. Drift velocity vs applied voltage for a superlattice embedded in a p-i-n diode showing NDV for large enough bias (from Ref. 60).

as described in detail in section 2.5.6.2. In spite of the narrowness of the miniband, all the main features of miniband conduction as discussed in the last sections are therefore present. Fig. 2.25 shows $V(F)$ relations computed with the help of the Boltzmann equation corresponding to the two SLs of Fig. 2.24. They show the very same general behavior as that of the experimental I-V characteristics as far as the temperature dependence is concerned: a decrease of the peak velocity (current) towards higher temperatures together with the shift of the peak voltage (critical field) and the reduction of the amplitude of the NDV (NDC). These observations are visually summarized in Figs. 2.26 and 2.27 together with a plot of the low-field mobility. A very good agreement between theory and experiment is therefore obtained. It could be most probably improved by a fully self-consistent treatment of the coupled transport-charge equations, which would correctly take into account field inhomogeneities and allow a direct plot of current-voltage rather than velocity-field quantities.

Particularly interesting is the temperature dependence of the low-field mobility, which follows rather well the prediction of the relaxation time approximation model (Fig. 2.26). At this point let us recall that this approximation is well-known to be valid in semiconductors in the low-field

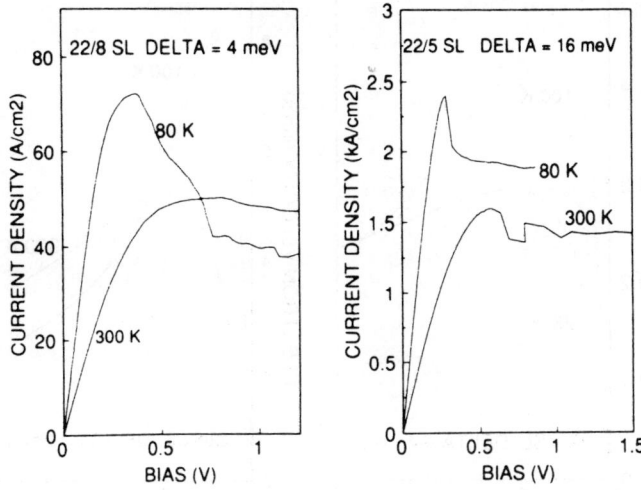

Fig. 2.24. Current-voltage characteristics of two narrow miniband superlattices (4 and 16 meV) exhibiting negative differential conductance (from Ref. 19).

limit, when only elastic collisions are considered. This agreement may therefore be regarded as an indication of two factors.

i) The most important contribution to the collision time is mostly *elastic* and temperature independent.

ii) Since ionized impurity scattering is a minor collision process in these moderately doped samples, interface roughness scattering appears to be the most probable mechanism limiting the mobility in these samples.

Remark i) is implicit in the various favorable comparisons with the relaxation time approximation model cited above. Remark ii) is comforted by the weak temperature dependence of interface roughness scattering.[21] Fig. 2.27 shows the temperature dependence of the transport properties of the narrowest miniband SL (4 meV) calculated by solving the Boltzmann equation incorporating interface roughness, impurity, and acoustical/optical phonon scattering. It is clear that the relaxation time approximation cannot be pushed too far, since a large deviation between the latter and the full Boltzmann equation is found as regards the peak velocity. In addition,

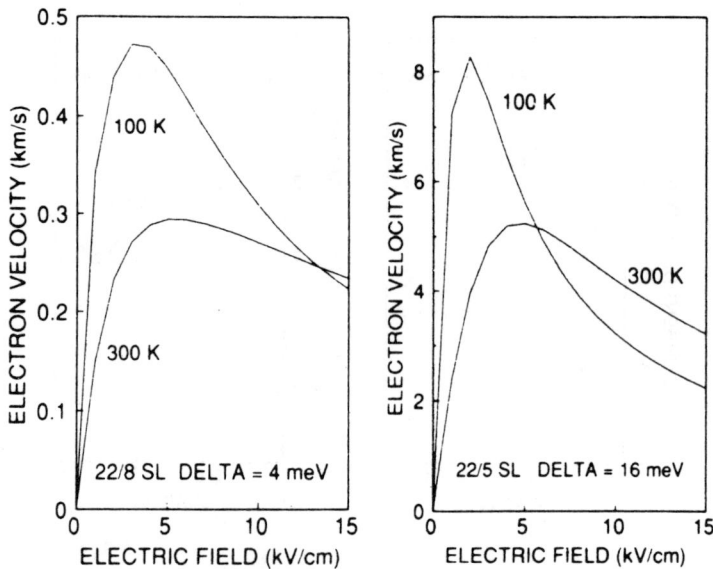

Fig. 2.25. $V(F)$ relations calculated by the Boltzmann equation for the two SLs of the previous figure (from Ref. 19).

the critical field is not constant between 77 and 300 K, as would be expected from the above considerations which stress the importance of temperature independent elastic scattering. It thus appears that phonon scattering plays a major role for large fields as intuitively expected, since temperature-dependent processes like phonon emission naturally prevail in a nonlinear regime.

We also note the saturation of the peak velocity towards low temperatures, an unexpected feature in the relaxation time approximation for narrow minibands. Indeed the temperature dependence in this model solely originates from the thermal broadening of the electronic distribution with respect to the miniband width. Unless for very low temperatures $T < \Delta/(2k_B)$, the factor $I_1(x)/I_0(x)$ with $x = \Delta/(2k_B T)$ indeed increases noticeably when T decreases. Heating through the applied electric field will tend to replace the *lattice* temperature T_L by the *electronic* temperature T_E in this expression. Since for large fields the latter is only marginally de-

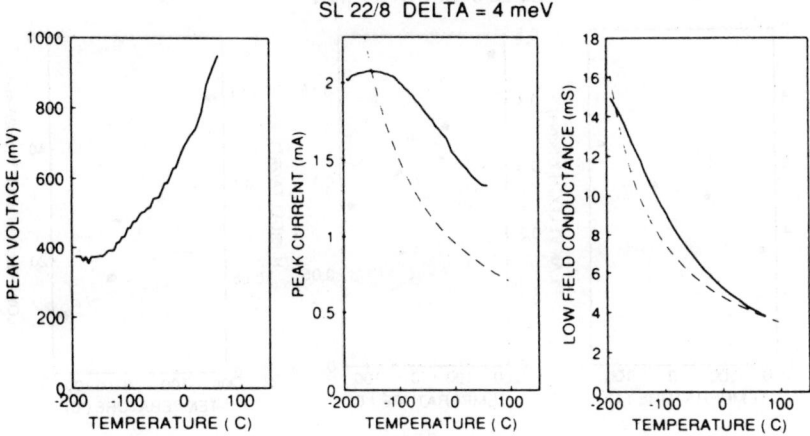

Fig. 2.26. Temperature dependence of the I-V characteristic in a 4 meV miniband width SL showing a good agreement with the relaxation time approximation model for the low-field mobility (from Ref. 19).

pendent on the lattice temperature for low T, we have a straightforward explanation for the small temperature dependence of V_p shown in Fig. 2.26. This interpretation has been confirmed by full Boltzmann equation calculations (Fig. 2.27) as well as by the balance equation approach mentioned above.[61]

In the spirit of Fig. 2.13, Fig. 2.28 shows a summary plot of μ/d^2, V_p/d, and edF_c in a large set of GaAs/AlAs superlattices with miniband widths Δ ranging from 4 to 130 meV including Grahn's data.[60] A very regular variation with Δ of both, the low-field mobility and the peak velocity, is demonstrated with an almost quadratic dependence for both over several orders of magnitude in the low Δ range. This behavior is actually acceptably described by the relaxation time approximation model, since Δ enters these two quantities twice, once in the Esaki-Tsu mobility and the second time in the factor I_1/I_0.

When a narrow miniband SL is highly doped, the interplay between charge and transport effects discussed in section 2.5.4 plays a strong role

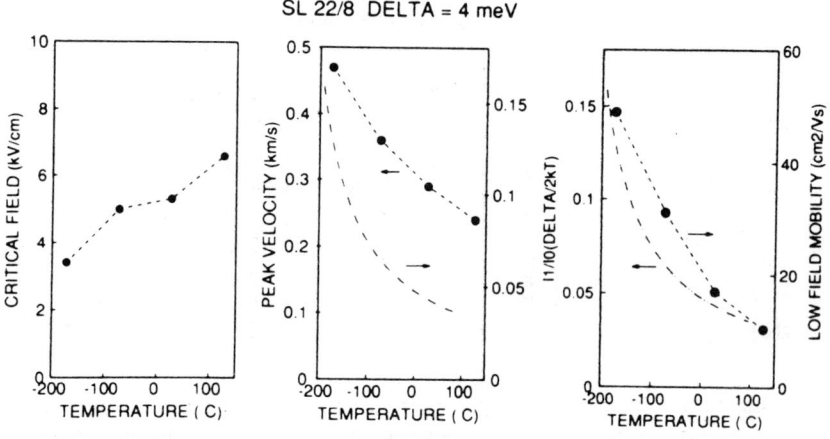

Fig. 2.27. Results of Boltzmann equation calculations for the SL of Fig. 2.26 (dashed line) compared to the relaxation time approximation model (dash-dotted line) (from Ref. 19).

down to the microscopic scale. This happens when the Debye length is of the order of the SL period, in which case local fields can vary largely within a single period. Obviously, such localized high-field domains will dramatically affect the SL conduction and I-V characteristics and are treated in detail in Chapter 5 of this volume.

2.5.3. *Miniband transport in the presence of disorder*

All the experimental results presented in the last section unequivocally show that narrow miniband superlattices indeed exhibit all the essential features of miniband conduction, down to miniband widths of a few meV much smaller than the thermal quantum $k_B T$. The consideration of disorder makes this conclusion somewhat puzzling. The above miniband conduction models indeed rely on the existence of delocalized Bloch states for the electrons with well defined wavevectors in order to apply the semiclassical dynamics of solid state physics within a band. In this picture any deviation of the crystal periodicity is treated as a weak perturbation and

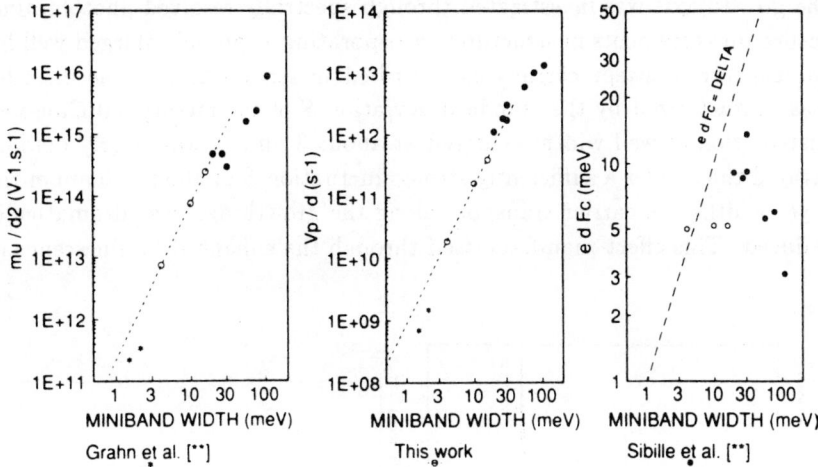

Fig. 2.28. μ/d^2, V_p/d, and edF_c vs miniband width in a large set of GaAs/AlGaAs superlattices (from Ref. 19).

taken into account as a phenomenon of scattering calculated to first order through the Born approximation. This is what is usually done in, e.g., the Boltzmann equation for homogeneous disorder like lattice vibrations or for inhomogeneous disorder like ionized impurity potential. However, when the lattice periodicity is too deeply disrupted, electron state localization by disorder may occur and the Born approximation fails.

Superlattices constitute an interesting case because they are strongly anisotropic. Since localization in one dimension is well-known to be more easily achieved than in two- or three-dimensional materials, one has to look keenly at structural SL disorder as a plausible reason of easy electron localization along the SL axis. It is common within the relaxation time approximation model to deduce scattering times of the order of 0.1 ps by fitting experimental data corresponding to a lifetime broadening of about 6 meV. This relatively large value indeed makes the very concept of a miniband, when it is narrow, questionable.

Interesting experiments in this spirit have been conducted by Cho-

mette et al.[62] who studied purposely disordered GaAs/AlGaAs superlattices by randomly altering the well widths during growth. Transport along
the growth axis was investigated through spectrally resolved photoluminescence measurements in structures incorporating a 'probe' enlarged well for
the capture of swept carriers as described in section 2.5.1. The disorder
was characterized by the standard deviation S of the intentional Gaussian
distribution of well widths centered at about 3 nm. It was clearly demonstrated that under a sufficiently strong fluctuation S of about 0.5 nm of the
layer widths the carrier transport along the growth axis was dramatically
reduced. This effect manifests itself through the enhanced luminescence of

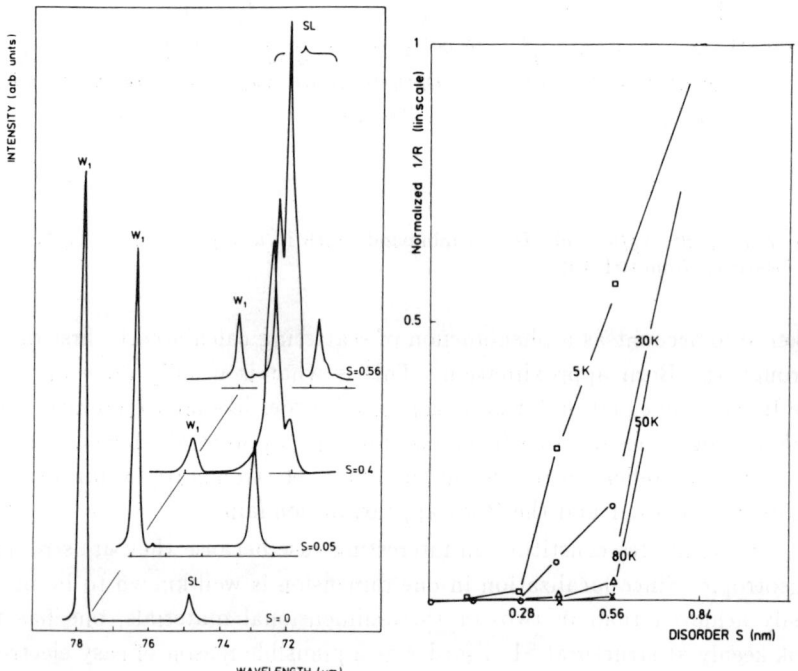

Fig. 2.29. Left: Photoluminescence spectra of four superlattices with increasing degree
of disorder S showing the increasing ratio of SL to enlarged well radiative efficiency
typical of a reduced photo-carrier mobility. Right: Temperature dependence of this
ratio for various values of the disorder parameter showing a preferential localization at
low temperatures (from Ref. 62).

the SL compared to the enlarged well implying a poor transport of photo-excited electron-hole pairs towards the enlarged well (Fig. 2.29).

The random well width variation is obviously responsible for fluc-tuations of the quantum state energies. For that reason miniband Bloch wavefunctions in an ideal superlattice are replaced by states which are loca-lized on a length scale defined by the magnitude of the disorder. However, this localization primarily affects states initially near the bottom or the top of the ideal miniband.

Indeed one locally thicker well will induce an energy level lower than the other superlattice states, which therefore will be mainly localized at the well site. Conversely, a narrower well will induce a higher energy level than the miniband top again essentially localized at the well site. Therefore, a close correlation between state localization and energy as depicted in Fig. 2.30 is present. We expect low energy states to be localized, and the existence of a *mobility gap* as in amorphous semiconductors. Beyond this gap, excited electrons will be available for the conduction, while they will be 'frozen' into these band-tail states below the mobility gap. Thermal excitation should induce carrier detrapping from the low-mobility, localized states and favor miniband conduction. This is exactly what the experiments of Chomette et al. indicated. For a given standard deviation of the well

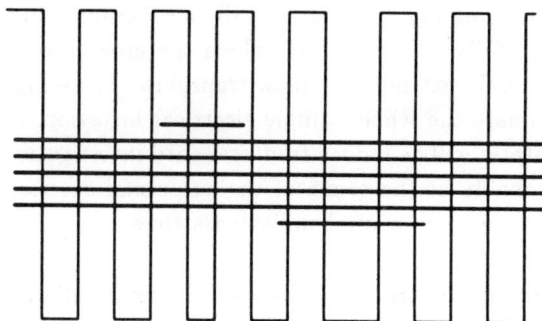

Fig. 2.30. Disorded superlattice with one localized state near the miniband top (narrower well) and one near the miniband bottom (thicker well).

thickness an increased transport along the SL axis at high temperatures was deduced from the smaller ratio of SL to probe well luminescence (Fig. 2.29). At elevated temperatures (80 K), there is a rather abrupt transition for a standard deviation approaching 2 monolayers (0.5 nm) beyond which the localization is strong even at this temperature. This effect is reminiscent of the famous Anderson transition in amorphous materials due to the interference effect of multiply scattered electron waves resulting in fully localized states. It appears from these results that a disorder of 2 monolayers standard deviation destroys a 70 meV wide miniband in reasonable agreement with rough estimations of the electron state energy fluctuations.[62] In this regime of strong disorder full electron state localization prohibits a calculation of the residual transport processes in terms of the Boltzmann equation. More appropriate is a description of transport as 'hopping' between localized states assisted, for instance, by electron-phonon coupling. This kind of scattering *favors* transport instead of inhibiting it, but the mobilities in this case are several orders of magnitude lower than through extended states.[63]

The artificially produced disorder in Chomette's experiments is particularly 'violent', since the whole SL is concerned irrespectively of the lateral degree of freedom. However, in naturally state-of-the-art grown structures, the well/barrier layer thicknesses fluctuate around a mean value, *randomly in the layer plane*. In other words, there is no or little long distance coherence of the disorder in the layer plane, because the size of the 'islands' defining one single well/barrier thickness is finite. For that reason in a SL sample of macroscopic size, there is generally a coexistence in various zones of the sample of fully localized states along the growth axis with others, which are sufficiently extended to allow transport. These high conduction paths will dominate the whole sample electrical behavior, in such a way that localized states will be hidden from standard investigations. This phenomenon is probably at the origin of the surprising success of miniband transport models in narrow miniband superlattices.

2.6. Miniband conduction vs Wannier-Stark localization and Bloch oscillations

Two old, major predictions of solid state physics -Wannier-Stark localization and Bloch oscillations- have been unambiguously verified in the last few years, both of them with superlattices as a test system. These fundamental phenomena are discussed in Chapter 3 of this volume. In this section we will be only concerned with their relation to miniband conduction.

2.6.1. *Miniband conduction vs Wannier-Stark localization*

The semiclassical miniband conduction models described in section 2.4 are based on scattering transitions between delocalized Bloch states *calculated in the absence of any electric field*. The only effect of the field is thus to modify the population of the various states. This approximation, although valid at small fields, may be expected to fail at large fields because the electron acceleration must be taken into account. In super-lattices, Wannier-Stark quantization is responsible for the breaking of the quasi-continuous miniband spectrum $E_z(k_z)$ in the growth direction into a ladder of discrete levels separated by eFd. When this spacing is less than the broadening $\hbar/\tau$ induced by the various scattering processes, the peaks in the one-dimensional SL density of states are washed out, and we can expect Wannier-Stark quantization effects to be negligible. However, when $eFd > \hbar/\tau$, Wannier-Stark localization should profoundly affect the transport properties.

As early as 1975, Tsu and Döhler have calculated transport in a Wannier-Stark ladder in terms of hopping induced by scattering between electron states localized by the electric field[64] (Fig. 2.31). Only acoustic-phonon scattering was considered, but the essential features of the calculation are believed to remain valid in the case of other elastic or inelastic scattering mechanisms. It was assumed that Fermi-Dirac statistics remains valid in the wells as the result of small hopping rates between the wells compared to the relaxation rates within a single well. The wavefunctions for finite fields were computed in the crystal momentum representation as a superposition of the miniband Bloch functions at $F = 0$. The current voltage characteristics show a linear behavior at low fields followed by a maximum at a critical field and negative differential conductance (Fig. 2.31). Because hopping is favored by scattering the conduction is higher at elevated than at low temperatures, i.e., an opposite behavior to miniband conduction. The original calculation apparently led to the conclusion that the critical field was a linear function of the miniband width Δ, again a characteristic in strong contrast to miniband conduction as described theoretically and experimentally in this chapter. Because this calculation is based on hopping between Wannier-Stark states, it is necessarily incorrect at low fields

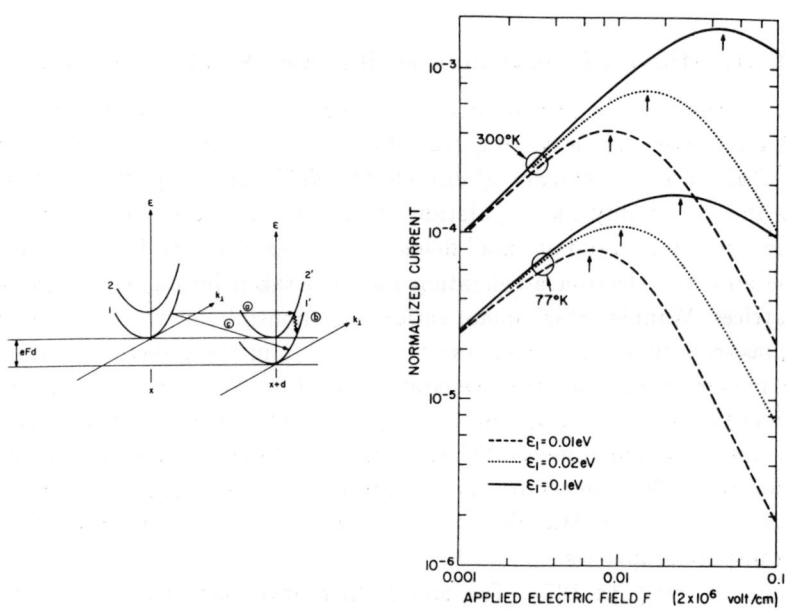

Fig. 2.31. Left: Tunneling between two Wannier-Stark states localized in adjacent wells and separated by eFd. Right: Dependence of the calculated current with the electric field for the tunneling between adjacent Wannier-Stark states (from Ref. 64).

when $eFd < \hbar/\tau$. Interestingly, the RHS of this inequality is exactly the critical field calculated semiclassically by Esaki and Tsu[3] (see section 2.4). In other words, a hopping transport model between Wannier-Stark states appears to be already valid *in the NDV regime predicted by the semiclassical models*. Although surprising, this conclusion is corroborated by a very simple analysis based on the Esaki-Tsu semiclassical model. Immediate algebra indeed yields

$$\frac{dV}{d\tau} = \frac{e \, \Delta \, d^2 \, F}{2 \, \hbar^2} \, \frac{1 - (F/F_c)^2}{[1 + (F/F_c)^2]^2} \qquad (2.21)$$

with $F_c = \hbar/(e\tau d)$. We deduce $dV/d\tau > 0$, if $F < F_c$, and $dV/d\tau < 0$, if $F > F_c$. In other words, *scattering inhibits transport when* $F < F_c$, *but enhances transport, when $F > F_c$*. We recognize in this behavior the signature of Bloch transport at low fields and hopping transport at high fields.

Implicit in the previous discussion is the conclusion that a Wannier-Stark ladder should be observable when $F > F_c$ and unobservable for F below F_c. This strong correlation between transport and the breaking of the miniband spectrum into Wannier-Stark resonances has been experimentally

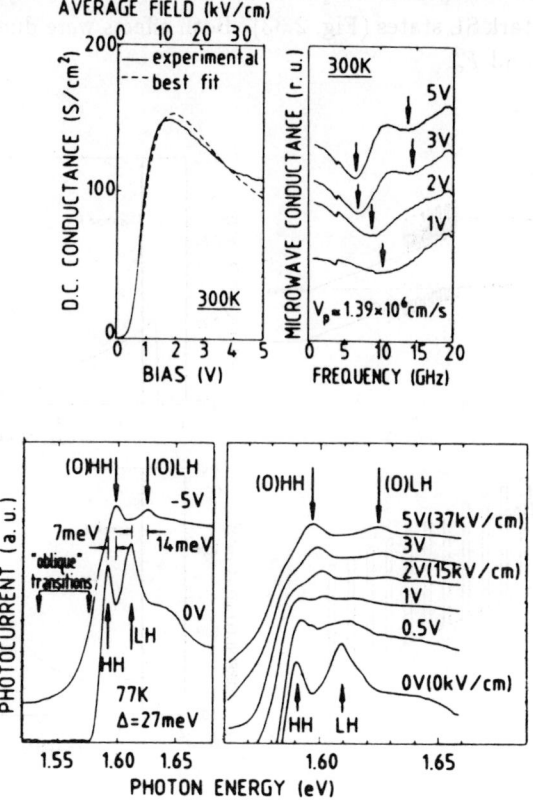

Fig. 2.32. Top: D.C. and microwave conductance of a superlattice demonstrating the existence of NDV. Bottom: Photocurrent spectra of the same sample demonstrating Wannier-Stark localization in the same bias range and the correlation between the two effects (from Ref. 65).

observed by comparing I-V data and photoconduction spectra *on the same sample.*[65] Both microwave NDV resonances and Wannier-Stark optical resonances were concurrently obtained in the very same field range (Fig. 2.32). This demonstrated unambiguously that the two phenomena are fundamentally correlated and are two different manifestations of the same basic physical phenomenon as argued above. The same kind of conclusion was reached by Beltram et al. using bipolar transistors with a superlattice located in the collector.[66] They observed NDC in the collector current characteristic of the transistor followed by a series of resonances under high collector bias, which could be interpreted as quantum reflection of the injected electrons by Wannier-Stark SL states (Fig. 2.33). Both effects were due to the biasing of the SL beyond F_c.

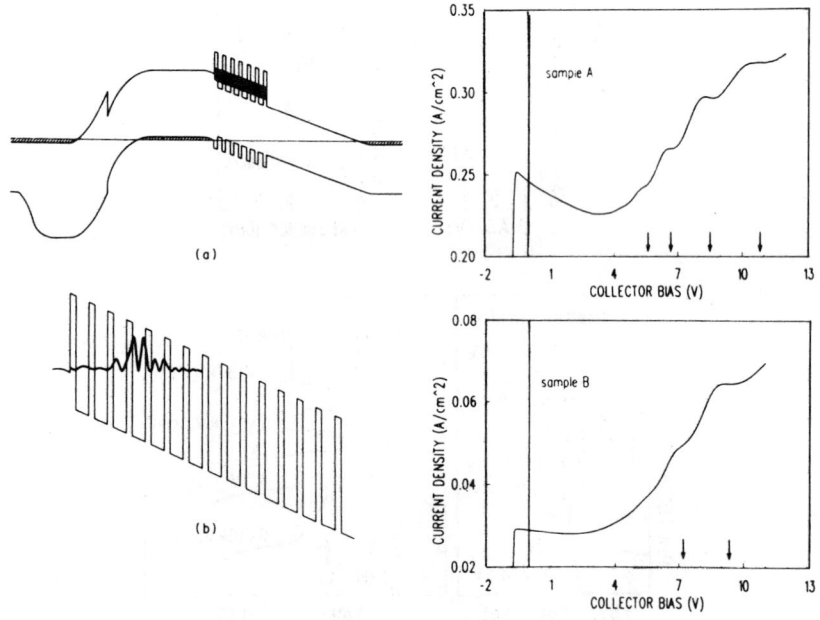

Fig. 2.33. Left: Energy-band diagram of a bipolar transistor with a superlattice in the collector intended to demonstrate quantum transmission resonances by the latter. The solid curve shows a typical electron wavefunction. Right: Collector current density vs collector bias under fixed emitter-current showing the superlattice resonances (from Ref. 66).

2.6.2. *Miniband conduction vs Bloch oscillations*

Bloch oscillations are at the root of NDV resulting from miniband conduction in a superlattice. It is indeed clear from the Esaki-Tsu model that electrons accelerated by the electric field will feel an increasing effective mass, which will eventually become negative, when $k > \pi/(2d)$ in the case of a sinusoidal miniband. For that reason the electron will finally slow down, until it accelerates again, and so on, resulting in a wavevector Bloch oscillation. The average electron momentum $< k >= eF\tau/\hbar$ equals $1/d$ when $F = F_c$, i.e., slightly less than $\pi/(2d)$. Therefore, it is clear that NDV indeed results from the rough completion of the Bloch oscillations for many of the accelerated electrons, but also that some of them will be scattered earlier because of the random character of collisions.

In the absence of scattering, *all* electrons would thus exhibit Bloch oscillations, the average velocity being consequently zero ($\tau = \infty$). The oscillations are, however, destroyed by collisions, and the average drift velocity increases with the field until the wavevector of the majority of the electrons exceeds $\pi/(2d)$ after which the velocity decreases again(i.e., NDV). This explanation draws a definite correlation between NDV, Bloch oscillations, and the negative effective mass of the dispersion relation responsible for the slowing down of electrons. However, these two last properties may not be directly related as shown by a simple application of the Esaki-Tsu model to a non-sinusoidal dispersion relation. Let us consider a hypothetical one-dimensional solid for which the energy dispersion would be $E(k) = \Delta[1 - cos(kd)]/2$, if $|k| < \pi/d$, and $E(k) = \Delta$ for $|k| > \pi/d$. In other words the Brillouin zone extends to infinity in this example case (or more precisely to π/a, a being the lattice parameter). There is therefore no super-periodicity, only the ordinary periodicity due to the atomic crystal structure is left. A critical wavector π/d has been artificially introduced, however, with the intention to mimic a superlattice minizone up to π/d beyond which the electron energy is constant. Since the superlattice does actually not exist, no Bloch oscillation may be expected unless for extreme fields, which would accelerate the electron up to $k \sim \pi/a$. It is now straightforward to show that by applying the Esaki-Tsu model to this bandstructure, we find NDV for a critical field and a peak velocity rather close to those obtained by periodizing the dispersion relation for $|k| > \pi/d$. Consequently, *negative effective mass is undoubtedly a sufficient driving force to be the cause of miniband NDV*, although in practice Bloch oscillations cannot be entirely discarded in ordinary superlattices possessing a near sinusoidal dispersion.

The crudeness of the Esaki-Tsu model is obviously inaccurate to estimate precisely the contribution of the two effects in a real case. Full solutions of the Boltzmann equation are therefore again required as carried out in[65] (Fig. 2.34). The relative population of electron states at the minizone center, the minizone boundary, and at $\pi/(2d)$ is shown in Fig. 2.34 as a function of the electric field. It is apparent that the fraction of electrons reaching the minizone boundary and thus being Bragg-diffracted is always smaller than those reaching the effective mass inversion point. Furthermore, there is almost a coincidence between the critical field, and the field, for which the population at $\pi/(2d)$ is maximum. These computations thus show that for a typical SL negative effective mass is the main cause of NDV, but it also shows that Bloch oscillations contribute to NDV in a somewhat smaller proportion.

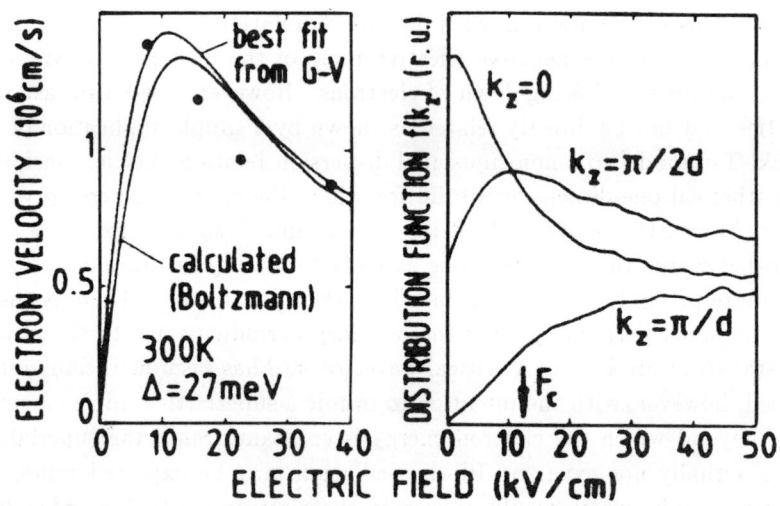

Fig. 2.34. Left: Experimental and calculated $V(F)$ relation for a GaAs/AlAs SL. Right: Electron distribution function calculated by the full resolution of the Boltzmann equation for three values of the electron momentum along the SL axis integrated over the in-plane wavevector. The computed critical field roughly coincides with the maximum of electron population at $\pi/(2d)$ (from Ref. 65).

2.7. Possible applications of miniband transport

The phenomena described above have been investigated mostly in the attempt of demonstrating and studying miniband conduction and its consequences. Potential applications to electronic and optical devices have not been really seriously considered to date. The author believes this situation may change in the near future due to the versatility in the design of microscopic structures, due to electro-optic coupling effects, and due to the short time scales involved in superlattice operation. One major advantage of SL or other quantum components is indeed the ability to entirely tailor the device properties by adjusting the microscopic parameters such as layer thicknesses and compositions. Secondly, the simultaneous manifestation of Wannier-Stark localization and negative differential velocity stresses the correlation between non-linear electronic transport and the electric field modulation of the optical absorption. Finally, the ultra-short sub-picosecond times involved in Bloch oscillations open the way to ultra-fast micro-optoelectronic effects which could be exploited in useful future applications. We shortly mention here two such possible applications making use of NDV at high frequencies.

2.7.1. *Millimeter/sub-millimeter oscillator sources*

The discussion in section 2.5.6.2 has highlighted the importance of high frequency resonances in the small-signal conduction of doped superlattices. Particularly interesting for applications is the high value of the negative differential conductance obtained at certain frequencies roughly integer multiples of $< V > /L$. By adequately scaling the microscopic and macroscopic parameters of the structure, very high frequencies can potentially be obtained. Taking 10^7 cm/s as a reasonably attainable peak velocity and $L = 0.5$ μm, we deduce a maximum fundamental frequency of 200 GHz eventually multiplied by an integer harmonic factor. Such high frequencies make superlattices excellent competitors to transferred electron devices or IMPATT diodes.

Because of their many common features, it is indeed fruitful to compare SL oscillators with Gunn effect based oscillators, which have been developed since thirty years ago up to 150 GHz for InP based devices. The main difference between the two devices is due to the microscopic origin of NDV based on intervalley transfer in the latter case. Among the several reasons for the origin of poor efficiencies in generated power at high frequencies, the ultimate limitation lies in the dynamics of accelerated electrons within the conduction band. The device operation indeed entirely rests on the periodic exchange of electrons between the main Γ valley and the satellite X or L valleys. However, this exchange is not instantaneous and demands collisions with phonons or other excitations to gain the momentum required by the transfer. In addition, it has been claimed by Kroemer that the energy relaxation of hot electrons within the Γ valley is a major bottleneck, which prohibits efficient power generation at frequencies larger than 100 GHz.[67] Because of the high intervalley energy gap, many phonons are indeed involved in the relaxation to the Γ minimum.

Superlattices may circumvent these problems, since the NDV is a phenomenon intrinsic to a single miniband, which is based on inherently fast Bloch oscillations. More precisely, the decrease of the velocity with the electric field begins, when the Bloch oscillation period becomes smaller than the mean collision time, which can be fully controlled by the magnitude of the applied electric field. For the sake of illustration we show in Fig. 2.35 calculations of the time response of the electron average velocity to an electric field step for different step heights. Bloch oscillations are clearly visible when $F > F_c$, and NDV can be observed in the stationary regime as well as at extremely short times. A fully self-consistent simulation accounting for quasi-ballistic transport in short SL structures characterized by inhomogeneous electric fields would clearly be necessary. However, the sub-picosecond time scale leaves us with the hope of ultra-high frequency (THz) operation of SL oscillators. There is indeed the old question to achieve the mythic *Bloch oscillator*, i.e., a solid state source based on Bloch oscillations to produce THz radiation. The direct observation of Bloch oscillations in the time domain is encouraging in this respect (see Ref. 68 and also Chapter 3 of this volume). However, the practical implementation of THz SL sources will need a tremendous reduction of parasitic extrinsic elements compared to Gunn effect sources, and considerable technological efforts will have to be carried out first.

At present the highest frequency of operation of superlattices has been reported by Hadjazi et al. in GaAs/AlAs SLs, who have demonstrated NDC up to 60 GHz limited by the instrumentation and sample processing.[69] Although this experiment only consisted of a small-signal measurement

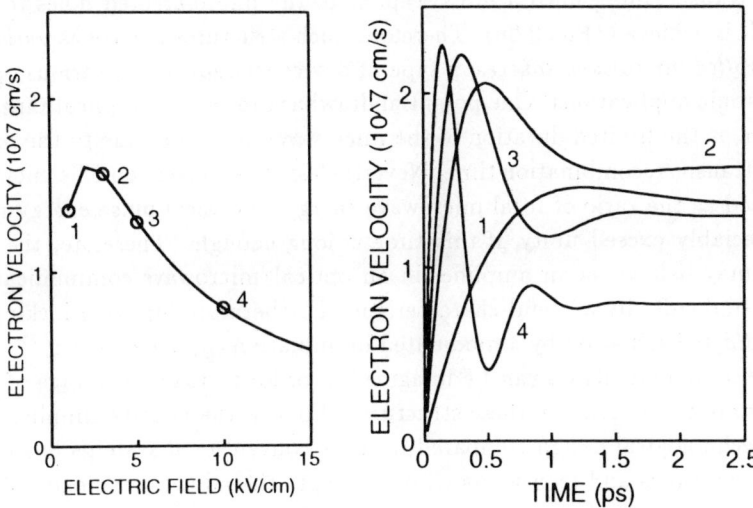

Fig. 2.35. Left: Stationary velocity-field relation for a GaAs/AlGaAs SL computed by the Boltzmann equation. Right: Time response of this SL to an electric field step from zero to its final value (shown by circles in the left graph). Subpicosecond overshoot and associated Bloch oscillations are clearly visible as soon as the electric field exceeds F_c.

and no source efficiency was determined, these results on non-optimized structures indeed tend to confirm the potentialities of superlattices in the millimeter wave range.

2.7.2. *Microwave/opto-electronic applications*

We have dicussed in section 2.5.6.3 the coupling between NDV miniband transport and the charge densities of carriers in n^+-p^--n^+ SL structures. This coupling may be at the origin of new effects demonstrated by Le Person et al. beyond the sole observation of NDV.[70] For a sufficiently large excitation and if holes are efficiently photo-created near the anode by suitable illumination, the field on this side is strongly reduced at the expense of the field in the rest of the SL which is enhanced. Because of NDV, the large concentration of secondary injected electrons builds up a moving high-field domain, which drifts along the SL up to the cathode, where it disappears. This process repeatedly occurs and induces a periodic photo-current response at a frequency $< V > /L$ equal to the inverse electron transit time , until a full sweepout of the photo-created holes to the anode is achieved (Fig. 2.36). Therefore, such structures behave as *optically controlled microwave sources*, a type of device susceptible to various optoelectronic applications. One potential drawback in certain applications, however, is the limited duration of the microwave operation due to the finite hole transit/recombination time. Nevertheless, the conversion efficiency expressed as the ratio of total microwave energy to optical pulse energy may appreciably exceed unity, if this time is long enough. Therefore, the device may behave as an amplifier in an optical/microwave communication link, undoudtedly a useful characteristic. Furthermore injection locking of the emitted radiation by a modulated illumination can be achieved [71], and more complex schemes can be imagined in order to take advantage of the electro-optic coupling in these structures. Finally, the relative simplicity of the technological design compared to more conventional systems based on photodetectors and transistors (if one excepts the added complexity of SL fabrication) contributes to make this type of device attractive.

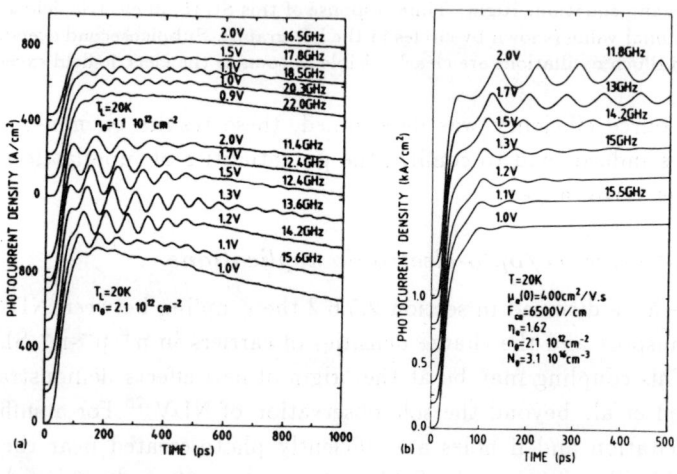

Fig. 2.36. Left: Experimental photoresponse of an illuminated superlattice for various applied biases showing high frequency current oscillations. Right: Simulated responses (from Ref. 70).

2.8. Conclusions

This chapter on miniband transport has not the ambition of covering all the various theoretical and experimental aspects of miniband transport. It is undoubtedly biased in the sense that some particular themes have been picked out and developed, mainly those which the author was most familiar with. One main message to be retained is the extreme diversity of effects based on superlattice miniband transport. Only a limited number of them, for which some experimental results were available, have been discussed here. Many others have been predicted on the basis of approximate models or sophisticated theories and await confirmation, if any. Superlattices are actually one excellent field of experimentation for testing some curious or novel predictions of quantum and solid state physics. As such it constitutes quite a rich field of investigation for the physicist. However, this utterly interesting object from the academic point of view has not yet proven to be very useful from a more practical application oriented point of view at the present time. Some hopes exist that it will allow a breakthrough in cases like THz frequency electronics/optoelectronics, if no other good solution exists at the frontiers of conventional semiconductor techniques. Let the future tell us, if this prospect will turn out to be true.

2.9. Acknowledgments

The present chapter would not have been written without the steady effort on superlattice transport carried out since many years by the Laboratoire de Bagneux of CNET, FRANCE-TELECOM, together with CNRS-LMM of Bagneux. The author has been one of the many actors involved in this research, of which the major initiative and stubborn confidence should actually be attributed to his excellent friend Jean-François PALMIER. Particularly important also was the participation of Christophe MINOT and Henri LE PERSON to this adventure. However, it would be unreasonable to omit many other contributions, which undoubtedly were nothing else than indispensable to the success of this research: among them I specially wish to thank Francis MOLLOT, Haila WANG, Richard PLANEL, Gohar ETEMADI, Mouloud HADJAZI, Eric DUTISSEUIL, François ALEXANDRE, Jean-Christophe HARMAND, Guy LE ROUX, Jean-Claude ESNAULT, and Sylvie VUYE. May the other contributors not be too angry with me, since only limited space prohibits me from naming them all...

References

1. L.V. Keldysh, *Sov. Phys. Solid State* **4**, 1658 (1963).
2. V.A. Yakovlev, *Sov. Phys. Solid State* **3**, 1442 (1962).
3. L. Esaki and R. Tsu, *IBM J. Res. Develop.* **14**, 61 (1970).
4. R. Dingle, W. Wiegmann, and C.H. Henry, *Phys. Rev. Lett.* **33**, 287 (1974).
5. L.L. Chang, L. Esaki, and R. Tsu, *Appl. Phys. Lett.* **24**, 593 (1974).
6. G. Bastard, *Wavemechanics applied to semiconductor heterostructures* (Les Editions de Physique, Les Ulis, France, 1988).
7. J.B. Krieger and G.J. Iafrate, *Phys. Rev. B* **33**, 5494 (1986).
8. R.H. Yan, R.J. Simes, H. Ribot, L.A. Coldren, and A.C. Gossard, *Appl. Phys. Lett.* **54**, 1549 (1989).
9. M.W. Peterson, J.A. Turner, C.A. Parsons, A.J. Nozik, D.J. Arent, C. Van Hoof, G. Borghs, R. Houdre, and H. Morkoç, *Appl. Phys. Lett.* **53**, 2666 (1988).
10. J.J. Song, Y.S. Soon, P.S. Jung, A. Fedotowsky, J.N. Schulman, C.W. Tu, and H. Morkoç, *Appl. Phys. Lett.* **50**, 1269 (1987).
11. B. Deveaud, A. Chomette, F. Clerot, A. Regreny, J.C. Maan, R. Romestain and G. Bastard, *Phys. Rev. B* **40**, 5802 (1989).
12. K. Fujiwara, K. Kawashima, T. Yamamoto, N. Sano, R. Cingolani, H.T. Grahn, and K. Ploog, *Phys. Rev. B* **49**, 1809 (1994).
13. T. Duffield, R. Bhat, M. Koza, F. DeRosa, D.M. Hwang, P. Grabbe, and S.J. Allen Jr., *Phys. Rev. Lett.* **56**, 2724 (1986).
14. G. Brozak, E.A. de Andrada e Silva, L.J. Sham, F. DeRosa, P. Miceli, S.A. Schwarz, J.P. Harbison, L.T. Florez, and S.J. Allen Jr., *Phys. Rev. Lett.* **64**, 471 (1990).
15. G. Belle, J.C. Maan, and G. Weimann, *Surf. Sci.* **170**, 611 (1986).
16. M. Helm, W. Hilber, T. Fromherz, F.M. Peeters, K. Alavi and R.N. Pathak, *Phys. Rev. B* **48**, 1601 (1993).
17. R.A. Suris and B.S. Shchamkhalova, *Sov. Phys. Semicond.* **18**, 738 (1984).
18. G. Brozak, M. Helm, F. DeRosa, C.H. Perry, M. Koza, R. Bhat, and S.J. Allen Jr., *Phys. Rev. Lett.* **64**, 3163 (1990).
19. A. Sibille, J.F. Palmier, M. Hadjazi, H. Wang, G. Etemadi, E. Dutisseuil, and F. Mollot, *Superlattices Microstruct.* **13**, 247 (1993).
20. M. Artaki and K. Hess, *Superlattices Microstruct.* **1**, 489 (1985).
21. A. Chomette and J.F. Palmier, *Solid State Commun.* **43**, 157 (1982).

22. I. Dharssi and P.N. Butcher, *J. Phys.: Condens. Matter* **2**, 4629 (1990).
23. J.F. Palmier, G. Etemadi, A. Sibille, M. Hadjazi, F. Mollot, and R. Planel, *Surf. Sci.* **267**, 574 (1992).
24. F. Capasso, K. Mohammed, A.Y. Cho, R. Hull, and A.L. Hutchinson, *Phys. Rev. Lett.* **55**, 1152 (1985).
25. F. Capasso, K. Mohammed, A.Y. Cho, R. Hull, and A.L. Hutchinson, *Appl. Phys. Lett.* **47**, 420 (1985).
26. A. Sibille, C. Minot, H. Le Person, J.F. Palmier, and F. Mollot, *Europhys. Lett.* **18**, 619 (1992).
27. H. Le Person, J.F. Palmier, C. Minot, J.C. Esnault, and F. Mollot, *Surf. Sci.* **228**, 441 (1990).
28. B. Deveaud, J. Shah, T.C. Damen, B. Lambert, and A. Regreny *Phys. Rev. Lett.* **58**, 2582 (1987).
29. B. Deveaud, J. Shah, T.C. Damen, B. Lambert, A. Chomette, and A. Regreny, *IEEE J. Quantum Electron.* **QE-24**, 1641 (1988).
30. B. Deveaud, A. Chomette, B. Lambert, A. Regreny, R. Romestain, and P. Edel, *Solid State Commun.* **57**, 885 (1986).
31. B. Lambert, B. Deveaud, A. Chomette, A. Regreny, and B. Sermage, *Semicond. Sci. Technol.* **4**, 513 (1989).
32. L. Esaki and L.L. Chang, *Phys. Rev. Lett.* **33**, 495 (1974).
33. J.F. Palmier, C. Minot, J.L. Lievin F. Alexandre, J.C. Harmand, J. Dangla, C. Dubon-Chevallier, and D. Ankri, *Appl. Phys. Lett.* **49**, 1260 (1986).
34. P. England, J.R. Hayes, E. Colas, and M. Helm, *Phys. Rev. Lett.* **63**, 1708 (1989).
35. A. Sibille, J.F. Palmier, H. Wang, and F. Mollot, *Phys. Rev. Lett.* **64**, 265 (1989).
36. A. Sibille, J.F. Palmier, C. Minot, and F. Mollot, *Appl. Phys. Lett.* **54**, 165 (1989).
37. P. Guéret, *Phys. Rev. Lett.* **5**, 256 (1971).
38. A. Sibille, J.F. Palmier, H. Wang, and F. Mollot, *Appl. Phys. Lett.* **56**, 256 (1990).
39. M. Büttiker and H. Thomas, *Z. Physik* **B 34**, 301 (1979).
40. A.A. Ignatov and V.I. Shashkin, *Sov. Phys. JETP* **66**, 526 (1987).
41. T.C.L.G. Sollner, *Phys. Rev. Lett.* **59**, 1622 (1987).
42. R.A. Suris and B.S. Shchamkhalova, *Sov. Phys. Semicond.* **24**, 1023 (1990).
43. B. Mitra and K.P. Ghatak, *phys. stat. sol. (b)* **164**, K13 (1991).

44. F. Aristone, A. Sibille, J.F. Palmier, D.K. Maude, J.C. Portal, and F. Mollot, *Physica* **B 184**, 246 (1993).
45. P.S. Kop'ev, R.A. Suris, I.N. Uraltsev, and A.M. Vasiliev, *Solid State Commun.* **72**, 401 (1989).
46. A. Sibille, J.F. Palmier, A. Celeste, J.C. Portal, and F. Mollot, *Europhys. Lett.* **13**, 279 (1990).
47. H. Noguchi and H. Sakaki, *Phys. Rev.* **B 45**, 12148 (1992).
48. J.F. Palmier, A. Sibille, G. Etemadi, A. Celeste, and J.C. Portal, *Semicond. Sci. Technol.* **7**, B283 (1992).
49. H. Noguchi, H. Sakaki, T. Takamasu, and N. Miura, *Semicond. Sci. Technol.* **9**, 778 (1994).
50. B.W. Hakki, *J. Appl. Phys.* **38**, 808 (1967).
51. J.F. Palmier (unpublished)
52. A. Sibille, J.F. Palmier, F. Mollot, H. Wang, and J.C. Esnault, *Phys. Rev.* **B 39**, 6272 (1989).
53. A. Sibille, J.F. Palmier, H. Wang, and J.C. Esnault, *Solid State Electron.* **32**, 1455 (1989).
54. G. Cohen and I. Bar-Joseph, *Phys. Rev.* **B 46**, 9857 (1992).
55. C. Minot, H. Le Person, J.F. Palmier, and F. Mollot, *Phys. Rev.* **B 47**, 10024 (1993).
56. I.B. Levinson and Ya. Yasevichyute, *Sov. Phys. JETP* **35**, 991 (1972).
57. X.L. Lei, N.J.M. Horing, and H.L. Cui, *Phys. Rev. Lett.* **66**, 3277 (1991).
58. Rolf R. Gerhardts, *Solid State Electron.* **37**, 681 (1994).
59. J.F. Palmier, *private communication*
60. H.T. Grahn, K. von Klitzing, K. Ploog, and G.H. Döhler, *Phys. Rev.* **B 43**, 12094 (1991).
61. X.L. Lei, *J. Phys.: Condens. Matter* **4**, L659 (1992).
62. A. Chomette, B. Deveaud, A. Regreny, and G. Bastard, *Phys. Rev. Lett.* **57**, 1464 (1986).
63. D. Calecki, J.F. Palmier, and A. Chomette, *J. Phys.* **C 17**, 5017 (1984).
64. R. Tsu and G. Döhler, *Phys. Rev.* **B 12**, 680 (1975).
65. A. Sibille, J.F. Palmier, M. Hadjazi, and H. Wang, in *Proceedings of the 21st International Conference on the Physics of Semiconductors*, edited by P. Jiang and H. Zheng (World Scientific, Singapore, 1992), p. 1190.

66. F. Beltram, F. Capasso, D.L. Sivco, A.L. Hutchinson, Sung-Nee G. Chu, and A.Y. Cho, *Phys. Rev. Lett.* **64**, 3167 (1990).
67. H. Kroemer, *Solid State Electron.* **21**, 61 (1978).
68. J. Feldmann, K. Leo, J. Shah, D.A.B. Miller, J.E. Cunningham, T. Meier, G. von Plessen, A. Schulze, P. Thomas, and S. Schmitt-Rink, *Phys. Rev.* *B* **46**, 7252 (1992).
69. M. Hadjazi, J.F. Palmier, A. Sibille, H. Wang, E. Paris, and F. Mollot, *Electron. Lett.* **29**, 648 (1993).
70. H. Le Person, C. Minot, L. Boni, J.F. Palmier, and F. Mollot, *Appl. Phys. Lett.* **60**, 2397 (1992).
71. J.F. Cadiou, J. Guena, E. Penard, P. Legaud, C. Minot, J.F. Palmier, H. Le Person and J.C. Harmand, *Electron. Lett.*, **30**, 1690 (1994).

CHAPTER 3

WANNIER-STARK LOCALIZATION AND BLOCH OSCILLATIONS

by FERNANDO AGULLÓ-RUEDA and JOCHEN FELDMANN

3.1. Introduction

This chapter discusses the effects of an electric field on the electronic and optical properties of semiconductor superlattices. Here an ideal superlattice is considered to be an infinite periodic series of strongly coupled quantum wells. The electric field is applied in the direction perpendicular to the layer planes.

From a stationary point of view the field reduces the interwell coupling and localizes the electronic states into a finite number of periods. At large fields this leads to the splitting of superlattice minibands into Wannier-Stark ladders. From a dynamic point of view electrons in the localized states describe very fast oscillations, called Bloch oscillations. These effects directly influence the static and dynamic optical properties of semiconductor superlattices producing interesting electro-optic effects.

The basic ideas of this chapter can be qualitatively understood with the help of Fig. 3.1. In the envelope-function approximation, which will be used throughout the chapter, the complexity of the electronic structure on the atomic scale is neglected.[1-4] Instead each layer is considered as a homogeneous medium with a certain effective masses and a conduction and valence band edge. Under these assumptions the electronic states can be calculated to a very good approximation by solving the Schrödinger equation with the appropriate matching conditions at the interfaces.

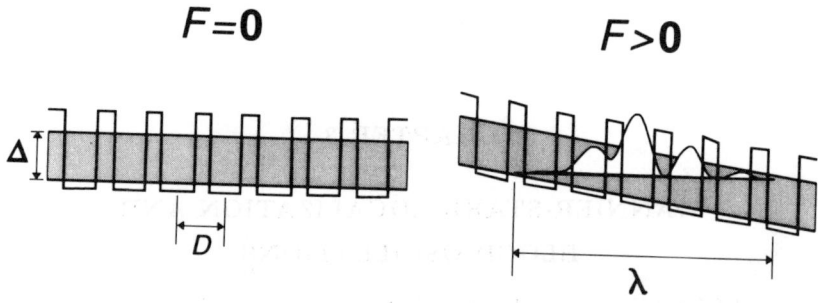

Fig. 3.1. Schematic representation of the effects of an electric field on the electronic properties of a semiconductor superlattice. The field is applied perpendicular to the layers.

At zero electric field (*flat band*) electron and hole levels form minibands with a certain dispersion relation due to the resonant coupling among all the well levels. The superlattice period is $D = L_W + L_B$, where L_W and L_B denote the quantum well and barrier thicknesses, respectively. The miniband width Δ depends on the interwell coupling and therefore on the period. Fig. 3.1 shows only the first miniband of the conduction band. In an ideal superlattice the levels of a miniband correspond to states that extend over the whole structure. The superlattice states can be thought as a superposition of the states of the individual wells mixed by the resonance between the well levels. In the same way electron states of a crystal can be considered as a superposition of the atomic states.

A constant electric field perpendicular to the layer planes (z-direction) introduces an electrostatic potential that detunes the interwell resonance and tilts the miniband. For a given energy level E (see Fig. 3.1) the corresponding state cannot extend beyond the miniband edges. This localizes the states within a distance $\lambda \approx \Delta/eF$ (*Wannier-Stark localization*[5,6]). The localization length λ approaches a single period at high fields. If the detuning between neighbor-well levels becomes larger than the intrinsic broadening, the miniband splits into a discrete series of levels (*Wannier-Stark ladder*). This means in the time domain that electrons are able to traverse the entire mini-Brillouin-zone without being scattered and thus perform *Bloch oscillations*.

The chapter is divided in two parts. The first part presents the Wannier-Stark localization and Wannier-Stark ladder formation, their observation in interband optical experiments, and some device applications. In the second part Bloch oscillations and their observation in optical four-wave mixing experiments are reviewed.

3.2. Wannier-Stark localization

3.2.1. *Historical overview*

3.2.1.1. *Stark ladder in the bulk*

The concept of the Wannier-Stark ladder appeared more than forty years ago in connection with the theoretical study of electronic bands in solids under electric fields. James[7] pointed out that an electric field should quantize the energies of electrons in a crystal band into a discrete levels separated by $\Delta E = eFD$. The first mathematical treatment of the Wannier-Stark states was given by Kane.[8] The term Stark ladder was introduced by Wannier.[9] To honor his contribution the name Wannier-Stark ladder is also used. A strong debate over the correct derivation of the Stark ladder states followed until recently. The experimental evidence was not clear enough because the electric fields needed to obtain a sufficient separation of the levels are difficult to attain.[10]

3.2.1.2. *Stark ladder in superlattices*

Semiconductor superlattices provided the ideal systems to probe these predictions. Their longer periods and narrower bands compared to bulk crystals made the localization length λ similar to the period D and brought Wannier-Stark localization into the real world.

The electric field effects on superlattices were treated first theoretically.[11] McIlroy[12] used a truncated superlattice of four wells to solve numerically the Schrödinger equation. He obtained a finite Stark ladder whose levels split linearly with the field at high fields and quadratically at low fields. He also obtained oscillations of the interband transitions. However, the fact that the number of wells was very small made it difficult to separate edge effects. Bleuse et al.[13] used a finite superlattice with many wells in a tight-binding approximation and predicted a blue shift of the absorption edge and oscillations of the absorption periodic in F^{-1} for a constant photon energy.

The first experimental observation of the Stark ladder formation in superlattices was reported by Mendez et al.[14] In photoluminescence and photocurrent (photoconductivity) experiments they clearly demonstrated the splitting of the optical transitions and the blue shift of the absorption edge. Photoconductivity was also shown to be the simplest technique to observe the phenomenon. Since then it has been extensively used. The oscillations periodic in F^{-1} were measured by Voisin et al.[15] in electrore-flectance experiments.

After these early experiments there has been a great deal of work on the Stark ladder in superlattices. Apart from the GaAs/AlGaAs system,[16-21] other materials like InGaAs-based superlattices[22,23] and GaAs/AlAs[24,25] have been investigated using other techniques like optical absorption,[22] resonant Raman scattering,[26,27] and differential photocurrent spectroscopy.[28] The Stark ladder formation has also served as a tool to measure the coherence length of the electron wavefunctions[16,29] or to induce doubly resonant Raman scattering by phonons.[26,27] The low field or strong coupling case case[17,20,21,24,31] and the transition from Franz-Keldysh to Stark ladder regimes[28,30-33] have been studied in detail.

Although these experiments were usually done at low temperatures in order to reduce scattering effects, the Wannier-Stark ladder has also been observed at room temperature,[17,29,34,35] where most applications are expected. Finally, the Stark ladder localization has been applied successfully to build some superlattice electro-optic devices, such as modulators[34,36-38] and self-electro-optic effect devices[34,35,39-41] (SEEDs). These devices are needed for fiber optics communications and optical computing.

3.2.2. *Properties at zero electric field*

The single-particle approximation considers one electron in an average electrostatic potential due to the ions of the crystal and the others carriers. No interaction with individual electrons and holes is taken into account. Therefore, excitonic effects are excluded. At zero electric field (*flat band*) the electrostatic potential is exactly periodic. All the single-well levels are at resonance and their coupling produces superlattice states that are completely delocalized over the whole superlattice. In a real superlattice, however, the periodicity is broken in part by impurities and fluctuations in the layer thicknesses and compositions and the extent of the wavefunctions is limited to a typical coherence length. Nevertheless, in good quality superlattices the coherence length can reach many periods.[16] As in the band theory for crystals, the superlattice levels form a miniband approximately centered at the single-well level E_1, e. g., the level of a well when there is no coupling with the others. Each superlattice state has an energy E and a momentum k_z perpendicular to the layer planes related by the dispersion relation $E(k_z)$.

In the tight-binding formalism the superlattice wavefunctions $\psi(k_z, z)$ are written as linear combinations of all the single-well wavefunctions $\phi(z)$. In the nearest-neighbor approximation the solution, which can be found in any solid-state textbook, gives approximate eigenenergies and eigenfunctions. The typical dispersion relation is

$$E(k_z) \approx E_1 - \frac{\Delta}{2} \cos(k_z D) , \qquad (3.1)$$

where E_1 is the single-well level measured from the bottom of the well. The miniband width is

$$\Delta \approx 4\gamma , \qquad (3.2)$$

where γ is the nearest-neighbor potential-overlap integral. The miniband widths of electrons are typically of a few tens of meV. For the heavy holes, which have a much larger effective mass than the electrons, the widths are of only a few meV. In this approximation the center of the miniband coincides exactly with E_1. The superlattice wavefunctions are

$$\psi(k_z, z) = \frac{1}{\sqrt{N}} \sum_n e^{ik_z nD} \phi(z - nD) . \qquad (3.3)$$

The wavefunctions have a complex structure with a superlattice envelope function modulating the fine structure coming from the single-well wavefunctions. It is interesting to note that the wavefunctions still have another fine structure at the atomic level, which is not taken into account in the envelope-function approximation. In a similar way, the existence of the quantum well structure can be discarded, when the movement of electrons in the perpendicular direction is considered as that of free electrons with an effective mass given by the curvature of the superlattice dispersion relation $E(k_z)$. However, this approximation breaks down at moderate electric fields when the miniband disappears. As we will see later, the superlattice envelope function is responsible for the Franz-Keldysh oscillations observed in the absorption spectrum and for the oscillator strength oscillations in the Stark ladder transitions. The exact dispersion relation and the electron wavefunctions can be calculated with the Kronig-Penney model. The center of the miniband appears slightly above the single-well level energy E_1.

The optical properties are determined by the density of states of electrons and holes, which can be obtained from the dispersion relations in the three spatial directions. Since electrons can move freely in the plane of the layers, the dispersion relations in that plane are those of the bulk material. For small k_{xy} they are quadratic (parabolic band approximation). In superlattices, due to the dispersion direction $E(k_z)$, the density of states is smoother than in uncoupled quantum wells and more similar to that in the bulk material.[42]

The absorption spectrum has two contributions. The first one corresponds to the excitation of unbound electron-hole pairs and is proportional to the joint density of states (JDOS) of electron and holes.[43] The absorption edge corresponds to the transition from the top of the heavy-hole miniband to the bottom of the conduction band. Its energy is $(\Delta_e + \Delta_h)/2$ lower than that of uncoupled quantum wells. The second contribution is due to the formation of excitons and produces excitonic peaks at the critical points of the density of states, namely the M_0 and the M_1 singularities, which are, respectively, at the bottom and at the top of the electron miniband.[44] In superlattices the electronic states extend over many periods, and the exciton binding energy is comparable to the one in the bulk. For a three-dimensional exciton the binding energy is four times smaller than for a two-dimensional exciton.[45] Therefore, excitonic peaks in superlattices are less pronounced than in uncoupled quantum wells.

3.2.3. *Theory*

In the single-particle approximation the movement of a carrier in the z-direction in an intrinsic superlattice is described by the following Schrödinger equation

$$\left(-\frac{\hbar^2}{2m^*}\frac{d^2}{dz^2} + eFz + U(z) \right)\psi(z) = E\psi(z) , \qquad (3.4)$$

where $U(z)$ denotes the superlattice potential and m^* the bulk effective mass (layer dependent). Due to the electrostatic energy an electric field introduces a separation $\Delta E = eFD$ between single-well levels. These levels are no longer exactly at resonance and the interwell coupling is reduced. If the separation is larger than the broadening of single well levels, the miniband can be resolved into a Wannier-Stark ladder. The ladder is formed by the levels of the N superlattice states, where N is the number of wells. For an infinite N, edge effects are absent and all the levels are evenly spaced. At high fields the levels of the Stark ladder coincide with the single-well levels.

Due to the special symmetry of the system, consecutive Stark ladder states have the same probability function shifted only by one period in space

$$|\psi_n(z)|^2 = |\psi_m[z - (n - m)D]|^2 . \qquad (3.5)$$

At a given electric field the superlattice states extend over the *localization length*

$$\lambda \approx \frac{\Delta}{e\,F} , \qquad (3.6)$$

which at low fields covers many periods. For fields of the order of $F \approx \Delta/(eD)$ the localization length approaches one superlattice period. This is referred to as the *complete localization* regime in the sense that the probability function is concentrated mainly in one well. However, it does not mean that an electron will stay in the well forever.

To perceive the full richness of the Stark localization one must solve the time-independent Schrödinger equation in the z-direction to obtain the Stark ladder states. These are not really bound, stationary states, but resonances in a continuum spectrum of energies. The potential energy of the electrons decreases in the direction opposite to the electric field. A single-well level is always in resonance with states of the continuum above the barriers. At low fields the resonance is small and the energy spectrum consists of evenly spaced sharp peaks (the Wannier-Stark ladder) on top of a continuum (see, for example, Ref. 46). The Stark ladder states are well confined in spatial regions inside the superlattice potential profile. The energy broadening Γ of the Stark ladder levels is given by[46]

$$\Gamma \approx \left(\frac{\hbar(e\,F)^2}{2\,m^*} \right)^{1/3} , \qquad (3.7)$$

where m^* is the bulk effective mass. For very high fields the broadening becomes larger than the barrier height and the Wannier-Stark ladder disappears. In this regime the superlattice states are strongly mixed with the continuum and the localization vanishes.

There are mainly two approaches to solve Eq. (3.4). One of the approaches[12,20,47-49] is to solve it numerically replacing the infinite superlattice by a finite number of wells. The potential $U(z)$ is considered as constant in the regions that are far apart from the quantum wells to get bound solutions. In each layer the wavefunction is written as a linear combination of Airy functions $Ai(z)$ and $Bi(z)$, and at the interfaces the Daniel-Ben Duke boundary conditions are used.[50] For a single quantum well it is easy to obtain an analytical solution, but for a large number of wells the problem is solved numerically with a transfer matrix method. Both the eigenvalues and the eigenfunctions can be found. The results are valid if the total length of the wells is much larger than the localization length λ at the corresponding electric field. For low electric fields the number of wells must be very large. In any case one must always take the solutions at the center regions to avoid edge effects.

In a simplified version[13,20] of this approach the potential is approximated by a piecewise profile. In each piece it is taken as constant and equal to the average value of the actual potential in the layer. Then in each piece the wavefunctions can be written as a linear combination of plane waves instead of Airy functions. This considerably simplifies the calculations. With one[13] or two[20] pieces per layer a rough idea of the Stark localization is obtained.

Figure 3.2 shows the superlattice wavefunctions calculated numerically for a 3.0 nm/3.0 nm-GaAs/$Al_{0.35}Ga_{0.65}As$ superlattice at various electric fields.[49] From the figure two conclusions can be drawn. First, heavy holes localize much faster than electrons and at 20 kV/cm they are almost completely localized in a single well. Second, at low fields the wavefunctions have several nodes to preserve the orthogonality between different wells. These nodes cause oscillations in the oscillator strength of the optical transitions that will be discussed later.

The other approach to solve Eq. (3.4) uses the tight-binding formalism[11,13,51-53] as it is done at zero field. This method gives a very intuitive picture of the Wannier-Stark localization. Superlattice wavefunctions $\psi_m(z)$ are written as linear combinations of all the single-well wavefunctions

F (kV/cm)

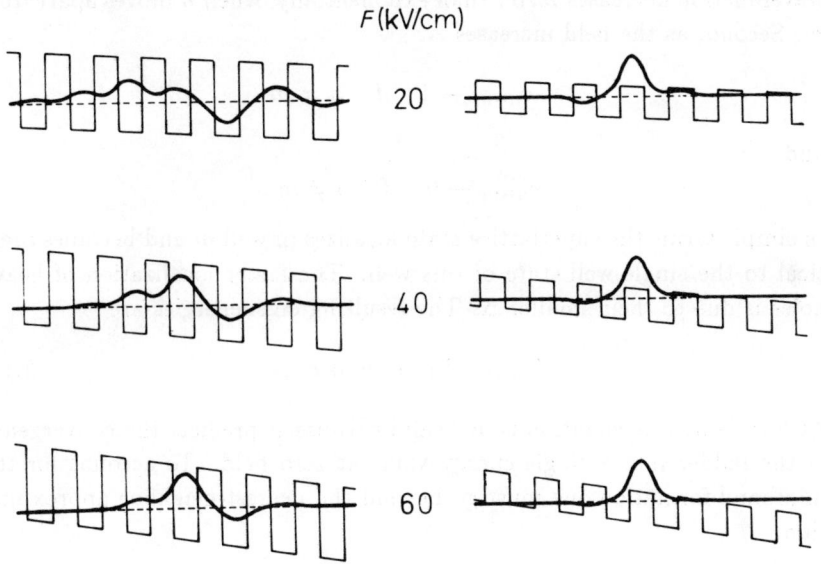

Fig. 3.2. Superlattice wavefunctions for electrons (*left*) and heavy holes (*right*) calculated numerically for a 3.0 nm/3.0 nm-GaAs/Al$_{0.35}$Ga$_{0.65}$As superlattice at various electric fields.[49] Only the states corresponding to the central well are shown (Adapted from Ref. 49.)

$\phi(z - nD)$

$$\psi_m(z) = \sum_n c_{n-m} \, \phi(z - nD) \, . \qquad (3.8)$$

Considering only nearest-neighbor overlap and neglecting coupling with other minibands, the coefficients c_{n-m} are given by[13]

$$c_{n-m} = J_{n-m} \left(\frac{\Delta}{2 \, e \, F \, D} \right) , \qquad (3.9)$$

where J_n denotes the Bessel function of the first kind of order n. For $\Delta \ll 2eFD$, the Bessel function can be approximated by[13]

$$J_{n-m} \left(\frac{\Delta}{2 \, e \, F \, D} \right) \approx \frac{1}{|n-m|!} \left(\frac{\Delta}{4 \, e \, F \, D} \right)^{|n-m|} , \qquad (3.10)$$

which shows mathematically the Wannier-Stark localization. First, the wavefunction decreases faster than exponentially, when n moves apart from m. Second, as the field increases

$$c_{n-m} \to 1 \quad \text{if} \quad n = m$$

and

$$c_{n-m} \to 0 \quad \text{if} \quad n \neq m .$$

In simple terms the superlattice state localizes in well m and becomes identical to the single-well state of this well. The faster localization of heavy holes is due to their smaller Δ. The resulting eigenenergies are

$$E_m = E_1 + m\, e\, F\, D . \tag{3.11}$$

At low fields this equation is not valid because it predicts the convergence of the ladder into a single energy value at zero field. To account for the miniband formation one must go beyond the nearest-neighbor approximation.

3.2.4. *Optical experiments*

The splitting of minibands and the localization of states have direct consequences for the optical properties. Because of the easy interpretation of the results, most experimental work on the Stark ladder in superlattices has been done with optical techniques.

In the experiments it is necessary to apply very high and uniform electric fields. For this purpose a long superlattice, containing of the order of one hundred periods in order to diminish edge effects, is grown in the intrinsic part of a p-i-n diode. The n layer is usually formed by the substrate, which is n-doped, and an intermediate layer grown by molecular beam epitaxy. After growth the sample can be etched down to the substrate to form a mesa structure with a diameter of a few hundred microns to reduce the dark current. Under a reverse bias voltage V ($V < 0$) the electric field in the intrinsic layer is approximately given by

$$F = \frac{(V_{bi} - V)}{W} , \tag{3.12}$$

where V_{bi} denotes the built-in voltage of the diode and W the intrinsic layer width.[54]

Among the optical techniques, photocurrent spectroscopy[55] is the best suited for moderate and high electric fields.[14,16] The sample is illuminated from the top side by monochromatic light. If the photon energy is above the superlattice absorption edge, it generates electron-hole pairs, which for finite fields are collected at the electrodes. The photocurrent spectrum, except for very special conditions,[56] resembles closely the absorption spectrum and has several advantages over other optical spectroscopies. First, it is easy to interpret, because each peak represents an optical transition. Second, it does not require the removal of opaque substrates as in absorption spectroscopy. Third, contrary to photoluminescence its sensitivity increases with the field, since electron-hole pair separation favors collection at the

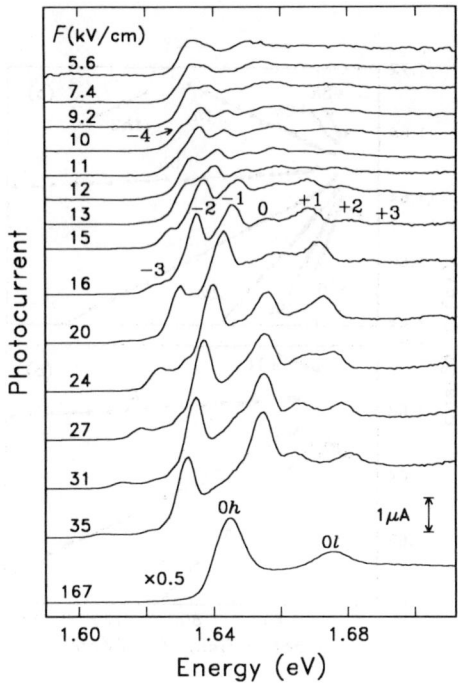

Fig. 3.3. Photocurrent spectra measured at 5 K for a 4.0 nm/2.0 nm-GaAs/ $Al_{0.35}Ga_{0.65}As$ superlattice at various electric fields. The peak labels refer to interwell indices (Adapted from Ref. 16.)

electrodes. Fourth, the sample itself acts as the light detector, producing a very high signal-to-noise ratio. Finally, fabrication of the electrodes does not imply extra work because they are already necessary in order to apply the electric field.

Figure 3.3 shows photocurrent spectra measured[16] at various electric fields in a 4.0 nm/2.0 nm-GaAs/Al$_{0.35}$Ga$_{0.65}$As superlattice at 5 K. At low electric fields the spectrum resembles the one at flat band with an excitonic peak for the transition from the top of the hole miniband to the bottom of the electron miniband. The peaks at 1.633 eV and at 1.652 eV correspond, respectively, to the heavy- and the light-hole transitions. At moderate fields the spectrum splits into a fan of peaks, which expands linearly with the field. This is clearly evidenced in Fig. 3.4, which represents the peak energies versus the electric field.

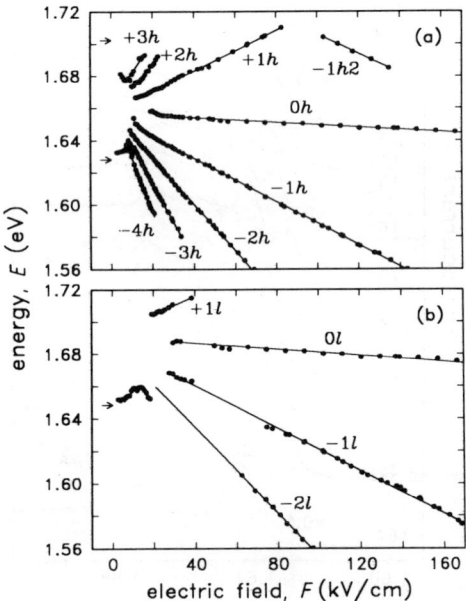

Fig. 3.4. Peak energies of photocurrent spectra plotted vs field measured at 5 K for a 4.0 nm/2.0 nm-GaAs/Al$_{0.35}$Ga$_{0.65}$As superlattice (see Fig. 3.3). Straight lines are least-squares fits to the data points in the linear segments. (a) Heavy-hole transitions. (b) Light-hole transitions (Adapted from Ref. 16.)

This a direct consequence of the splitting of the minibands into Stark ladders. Each hole state n overlaps with electron states m, producing as many optical transitions labeled $(m - n)$ (Fig. 3.5). The letters h or l are added to identify, respectively, heavy- or light-hole transitions. If $n = m$ the hole and the electron states correspond to the same well and the transition is called *intrawell* to distinguish it from *interwell* transitions between states corresponding to different wells ($n \neq m$). When the interwell coupling is small, transition energies can be obtained from Eq. (3.11). If the electric field is constant throughout the superlattice, transitions with the same $p = m - n$ are identical and the energies depend only on the separation between the corresponding wells

$$E_p(F) = E_0(F) + p \, e \, F \, D \, . \tag{3.13}$$

$E_0(F)$ is the intrawell ($p = 0$) transition energy

$$E_0(F) = E_g + E_{1e}(F) + E_{1h}(F) \, , \tag{3.14}$$

where E_g is the bandgap of the well bulk semiconductor. The field dependence of $E_0(F)$ is due to the Stark effect in a single quantum well.[57]

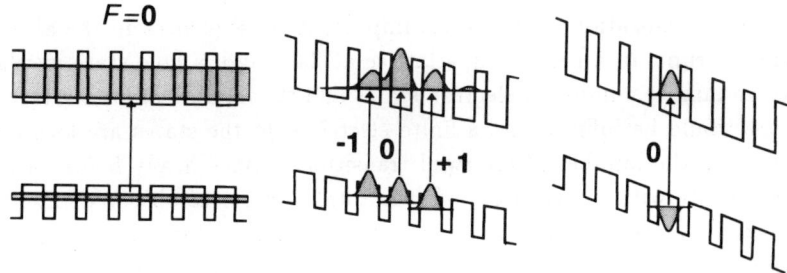

Fig. 3.5. Sketch showing the origin of interwell transitions between different Stark ladder states. The labels give the index p.

In second-order perturbation theory[58]

$$E_0(F) = E_0(0) - 2.88 \times 10^{-11} \frac{(m_e^* + m_h^*)}{m_0} L_W^4 F^2, \qquad (3.15)$$

where m_e^* and m_h^* denote, respectively, the electron and the hole effective masses and m_0 the free electron mass. This produces a red shift in the optical transitions, which depends quadratically on the electric field and is mainly due to the contribution of the heavy-hole, since it has a much larger mass. The shift can be seen in the experiments at high fields for the intrawell transitions $0h$ and $0l$ (Fig. 3.3).

Superlattices, in which the Wannier-Stark ladder is observed, have narrow wells in order to achieve a strong coupling. Consequently, the Stark shift is very small, and E_0 is almost field independent. The interwell transition energies move linearly with the field as it is observed experimentally (Fig. 3.4). The energies plotted in this figure are really those of excitonic peaks. Excitonic effects introduce small corrections to the single-particle approximation of Eq. (3.13) as it will be discussed later.

According to Eq. (3.13) the slopes of the transition energies vs field depend linearly on the index p and on the superlattice period D. This has been tested experimentally[16] for different indices p and for different superlattice periods (see Fig. 3.6).

3.2.5. *Localization length and coherence*

The localization of states has important consequences in the absorption spectrum. In a perfect superlattice at flat band, a hole state overlaps with an infinite number of electron states and the number of interwell transitions would be infinite. At a finite electric field the states are localized, decreasing the number of interwell transitions. Since heavy holes localize very quickly with increasing field, the total number of observed interwell transitions gives directly the extent of the electron states in superlattice periods (Fig. 3.5). Therefore, the intensity of interwell transitions easily allows a mapping of the electron Stark-ladder states. The intensity is proportional to the overlap of electron and hole states. When holes are completely localized, it is proportional to the area under the electron squared wavefunction contained in the corresponding well.

In a real superlattice the localization length is limited by the coherence length at zero field. The spatial extent or *coherence length* of the super-

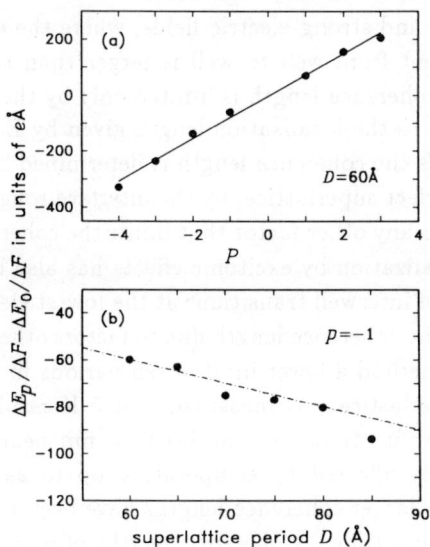

Fig. 3.6. (a) Slopes of the fitted straight lines in Fig. 3.4 as a function of the index p. (b) Slope of the transition $-1h$ vs superlattice period measured for different superlattices. The dashed-dotted line represents the dependence expected from Eq. (3.13), if E_0 is assumed to be constant (Adapted from Ref. 16.)

lattice states is determined in the first place by deviations from the ideal periodicity. These deviations produce energy differences between single-well levels and reduce the interwell coupling. The consequence is a localization of the states and a decrease of the coherence length. As we have seen the periodicity can be broken by an electric field. However, it can occur also by any disorder in the superlattice due to fluctuations in the growth parameters such as layer widths and composition. The effects on the coherence are similar in both cases, but the electric field allows to do it in a controlled way.

The coherence length depends on the differences between single-well energy levels. If they are much smaller than the miniband width then the coherence can extend over many periods. This is the case for strongly coupled superlattices close to flat band. If the differences are of the order or larger than the miniband width, the coherence is limited to one period.

This is the case for superlattices with very wide barriers (multiple quantum wells) or for superlattices at very high fields.

For moderate and strong electric fields, where the electric-field induced energy increment from well to well is larger than the one associated with disorder, the coherence length is limited only by the field strength. In this case it is equal to the localization length given by Eq. (3.6). However, for low electric fields the coherence length is determined by the superlattice disorder or, in a perfect superlattice, by the interface roughness, impurities, phonons, alloying or any other factor that limits the coherence length in the bulk material. Localization by excitonic effects has also been reported.[59]

The number of interwell transitions at the lowest electric fields gives a lower estimate for the coherence length due to factors other than the electric field.[16] With this method a lower limit of ten periods in a 4.0 nm/2.0 nm-GaAs/AlGaAs superlattice was measured[16] at 5 K resulting in an extent of the electron wavefunctions over at least 60 nm near flat band. This limit is only slightly affected by temperature up to 292 K.[29] In shorter period superlattices larger coherence lengths have been observed[29] because the miniband becomes wider. Coherence lengths of more than 90 nm have been reported.[29] It seems that this lower estimates are still well below the true values of the coherence length.

At fields of the order of $F \approx \Delta/eD$, where complete localization occurs, interwell transitions with $p \neq 0$ disappear and only the intrawell transition ($p = 0$) remains (Fig. 3.3). The latter determines the absorption edge. At these fields Stark ladder levels coincide with the single well levels. Since these levels lie slightly below the miniband center and the heavy-hole miniband is very narrow, the high-field absorption edge is blue shifted by about $\Delta_e/2$ with respect to the zero-field absorption edge. This can be seen by comparing the uppermost and the lowest spectra in Fig. 3.3. The blue shift, which opposes the red shift found in uncoupled quantum wells, has applications for light modulators and other electro-optic devices, which will be discussed later.

The absorption spectrum at very high fields looks very much like the spectrum of uncoupled wells with similar transitions energies and excitonic features. The latter are broader due to the tunneling of electrons out of the well. At intermediate fields the absorption edge is smeared out due to the presence of interwell transitions. In the tight-binding approximation the absorption coefficient at photon energy E is given by

$$\alpha(E) = \sum_{p=-N}^{N} \alpha_p .\qquad(3.16)$$

α_p is the absorption[13] due to the interwell transition p

$$\alpha_p(E) = (2N+1)\alpha_0 J_p^2 \left(\frac{\Delta_e + \Delta_h}{2eFD}\right) Y[E - (E_g + E_{1e} + E_{1h} + peFD)] ,\qquad(3.17)$$

where E_g denotes the band gap of the well material and $Y(x)$ the unit step function [$Y(x) = 1$ for $x > 0$ and $Y(x) = 0$ otherwise]. The indices e and h refer to electrons and holes, respectively. For a GaAs/AlGaAs superlattice[13] $\alpha_0 \approx 0.006$. Calculated absorption spectra are given in

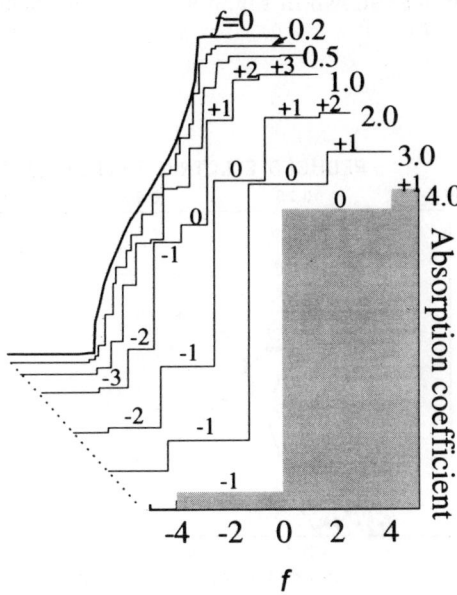

Fig. 3.7. Absorption spectra for a GaAs/AlGaAs superlattice as a function of the electric field calculated in the tight-binding formalism. $\epsilon = E_0 - 4(E_g + E_{1h} + E_{1e})/(\Delta_e + \Delta_h)$ and $f = 4eFD/(\Delta_e + \Delta_h)$ are, respectively, the reduced photon energy and the reduced electric field (Adapted from Ref. 13.)

Fig. 3.7 for a typical superlattice as a function of the applied electric field. The figure shows the same behavior as the experimental spectra of Fig. 3.3 except for excitonic effects. At moderate fields the spectra split into inter-well transitions which expand linearly and become weaker with field. At high fields their disappearance produces a blue shift of the absorption edge compared to flat band.

The oscillator strength I_p of interwell transition p is proportional to the square of the Bessel function. At low and moderate electric fields the intensity has strong oscillations, due to the overlap of the various nodes of electron and hole Stark ladder wavefunctions. These oscillations have been observed experimentally[20,29] and are shown in Fig. 3.8. Exciton and single-particle oscillator strengths are roughly proportional.

Transitions with $p < 0$ are generally weaker than those with $p > 0$. This asymmetry is not explained by the tight-binding approximation. Numerical calculations[20,49] have correctly given the oscillator strength and the asymmetry on p as shown in Fig. 3.9. Excitonic effects further enhance the asymmetry.[60-62]

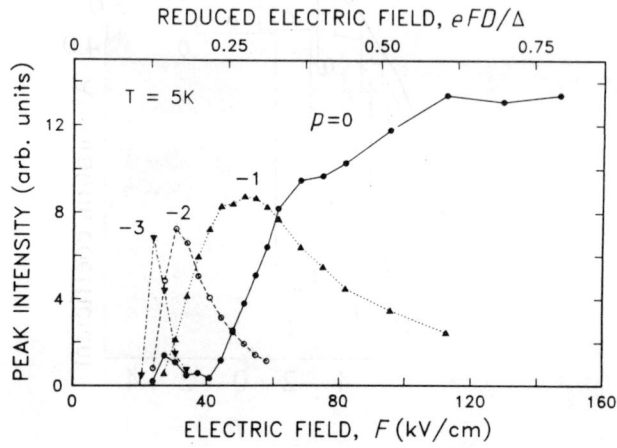

Fig. 3.8. Photocurrent intensity of interwell transitions (with indices p) as a function of the electric field measured at 5 K in a 4.0 nm/1.5 nm-GaAs/Al$_{0.3}$Ga$_{0.7}$As superlattice (Adapted from Ref. 29.)

For fields $F \gg \Delta/2eD$ the asymptotic approximation for the Bessel function [Eq. (3.10)] leads to a very simple relation between the oscillator strengths of interwell transitions

$$\frac{I_p}{I_0} \approx \frac{1}{(p!)^2} \left(\frac{\Delta}{4\,e\,F\,D}\right)^{2p} . \tag{3.18}$$

This equation gives a linear dependence with $(eDF)^{-2p}$ that has been observed experimentally.[24,63] The slopes depend on the superlattice period as reflected in Fig. 3.10 due to the presence of Δ in Eq. (3.18).

Finally, at very high fields the limit of the Bessel functions gives $I_p \to 0$ for $p \neq 0$, which expresses mathematically the complete localization of states and the disappearance of interwell transitions with $p \neq 0$.

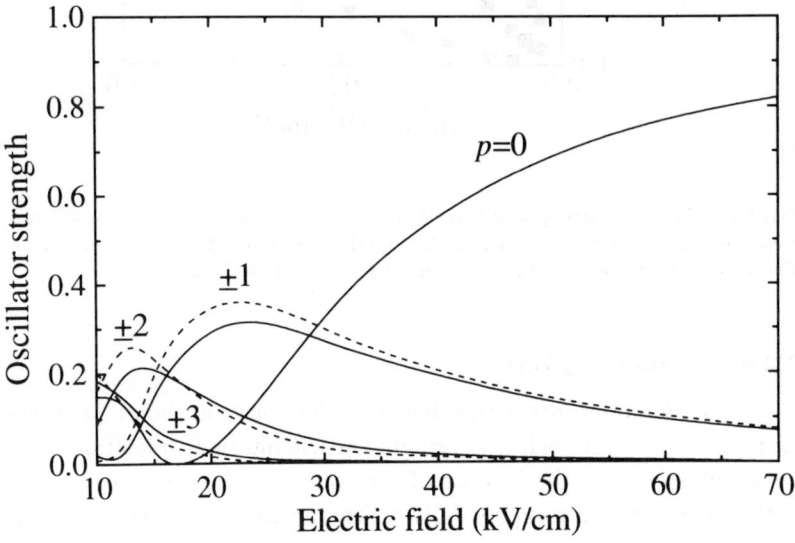

Fig. 3.9. Calculated oscillator strength of interwell transitions as a function of electric field in a 3.0 nm/3.0 nm-GaAs/Al$_{0.35}$Ga$_{0.65}$As superlattice. Dashed curves are for $p < 0$ (Adapted from Ref. 49.)

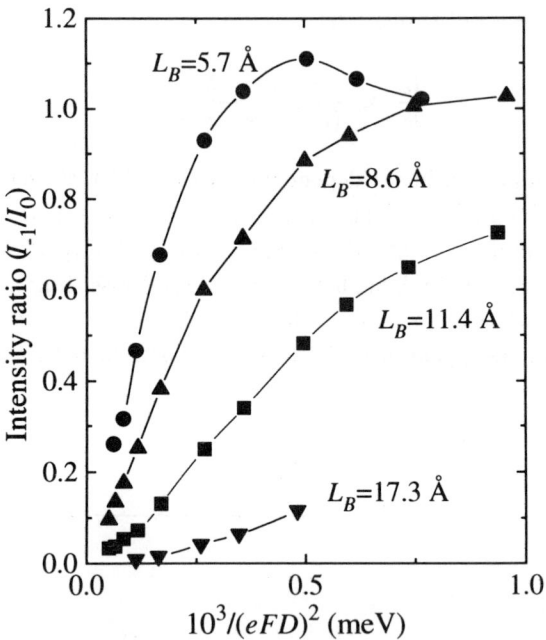

Fig. 3.10. Oscillator strength of transition $p = -1$ normalized by the transition $p = 0$ measured vs electric field in GaAs/AlAs superlattices with different barrier widths L_B. The well width was always $L_W = 3.1$ nm (Adapted from Ref. 62).

3.2.6. *Excitonic effects*

The Coulomb interaction between electrons and holes produces an exciton peak slightly below the single-particle interwell transition E_p. The peak corresponds to the $1s$ state of the exciton, and its separation from E_p is the exciton binding energy $E_p^b(F)$. The peaks from other hydrogen-like states ($2s$ and so on) are too weak to be seen.

The interwell exciton transition energies $E_p^x(F)$ are given by Eq. (3.13) modified to include the exciton binding energy $E_b^p(F)$

$$E_p^x(F) = E_0(F) - E_b^p(F) + p\,e\,F\,D. \qquad (3.19)$$

At zero electric field the binding energy and the size of excitons in strongly coupled superlattices[64] approach the bulk values because the states extend over many periods (for GaAs the heavy-hole exciton binding energy and the Bohr radius[65] are, respectively, 4.7 meV and 12 nm). The behavior with electric field is different for intrawell and interwell excitons.

For intrawell excitons, when electron states are localized either by increasing the superlattice period[64] or by increasing the electric field,[19] the binding energy increases. At high fields the exciton becomes quasi two-dimensional with the electron and the hole confined in the same layer, and its binding energy approaches that of an uncoupled quantum well. The transition from a three-dimensional to a quasi-two dimensional exciton should occur for the field where the exciton is squeezed beyond its Bohr diameter. The heavy-hole exciton binding energy measured for a 4.0 nm/4.0 nm-GaAs/AlGaAs superlattice[19] changes sharply from less than 8 meV at low fields to about 12 meV for fields above 20 kV/cm. The change in the binding energy produces a small, but noticeable nonlinear decrease in the intrawell excitonic transition at low fields.[16] This can be seen in Fig. 3.4 for fields around 25 kV/cm.

The influence of the field on interwell excitons has not been measured yet. Intuitively the binding energy should decrease at high fields because the electron and the hole tend to confine in different wells. The high field binding energy should be smaller for larger values of p as the average electron-hole separation gets larger.

The exciton effects have been calculated in detail as a function of the field by several authors.[48,49,53,60−62,66] The influence of the superlattice period has also been considered.[62,66] Fig. 3.11 shows the binding energy calculated for a typical superlattice. From the figure several conclusions can be drawn. As expected, for the intrawell exciton at high fields the binding energy increases.[19,48] At low fields it oscillates with field due to the nodes in the Stark ladder wavefunctions in much the same way as the single-particle oscillator strength.

For interwell excitons the binding energy first reaches a maximum and then decreases slowly down to a constant value. This value corresponds to the Coulomb interaction between an electron and a hole moving in planes separated by p periods.[48] The maximum occurs for the field where the probability of finding both carriers in the same well is larger.[49] This field decreases as $|p|$ increases. This is possible only at moderate fields when the superlattice envelope function has a wavy shape and peaks at wells other

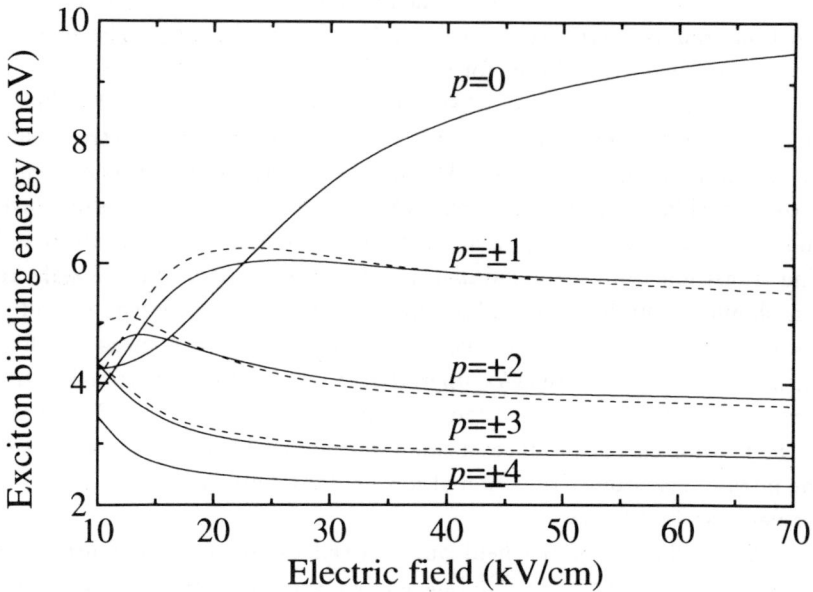

Fig. 3.11. Binding energy for interwell excitons calculated as a function of electric field in a 3.0 nm/3.0 nm-GaAs/Al$_{0.35}$Ga$_{0.65}$As superlattice. Dashed curves are for $p < 0$ (Adapted from Ref. 49).

than the central one. As discussed before close to the maximum the oscillator strength is larger for excitonic transitions with $p < 0$.[49] This is due in part to the asymmetry in the overlap of the electron and hole single-particle wavefunctions and to pure excitonic effects.[60,62] The binding energy also shows some asymmetry, but its sign depends on the interwell coupling. For long period superlattices the binding energy is larger for excitons with $p < 0$, whereas for short period superlattices it is the opposite.[62] The latter case reflects the situation in Fig. 3.11. Both the exciton oscillator strength and the binding energy present some oscillatory structure at very low fields due to the complex structure of the single-particle wavefunctions near flat band.[62] These oscillations have been observed experimentally (Fig. 3.8).

The absorption coefficient at constant photon energy oscillates when the electric field is swept.[15,16,18,29] The oscillations are due to the resonance of the photon energy with the different interwell transition energies. It can be easily understood from Fig. 3.4. A photon energy below the flat band absorption edge crosses all the interwell transitions with $p \neq 0$. The crossing points define the electric fields for which the absorption at constant energy has a maximum. The peaks are mainly due to excitonic effects. Interwell transitions with large p move very rapidly with field, and it is better to use the intensity vs field than the spectra to measure their energies. For energies above the flat band absorption edge the oscillations are weaker because of the large absorption background. Fig. 3.12 shows the oscillations of the photocurrent vs electric field measured[16] at different photon energies in a GaAs/AlGaAs superlattice. Between the peaks the *I-V* characteristic of the superlattice p-i-n diode shows zones of negative differential resistance, which are very interesting for bistable electro-optic devices.

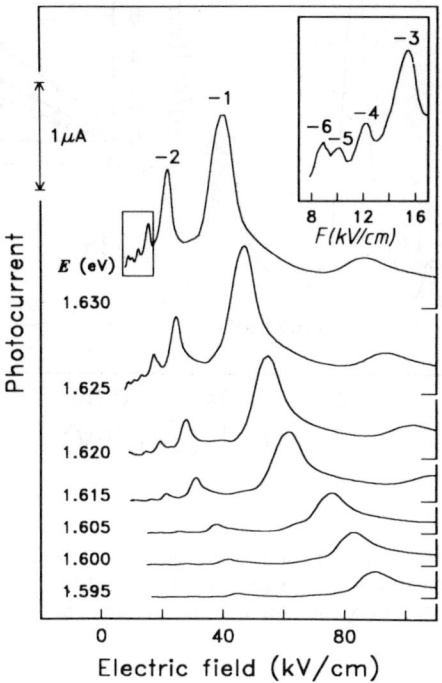

Fig. 3.12. Photocurrent intensity versus applied electric field measured[18] at different photon energies in a 4.0 nm/2.0 nm-GaAs/AlGaAs superlattice. The peak labels refer to heavy-hole interwell transition index p.

The oscillations are periodic in F^{-1}, resembling the Shubnikov-de Haas oscillations in semiconductors under magnetic fields. The periodicity can be understood by rewriting Eq. (3.13) as

$$F_p^{-1} = p\, e\, D\, [E(F) - E(0)]. \qquad (3.20)$$

The F^{-1} dependence is observed experimentally, as shown in Fig. 3.13.

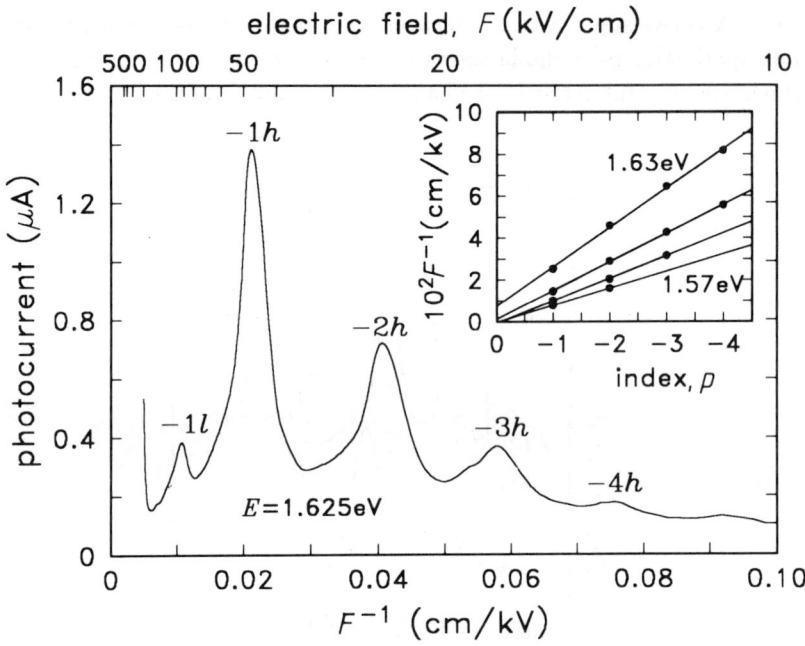

Fig. 3.13. Photocurrent vs reciprocal electric field in a 4.0 nm/2.0 nm-GaAs/AlGaAs superlattice at 5 K measured[16] for a photon energy of 1.625 eV. The inset shows the linear dependence of the maxima with F^{-1} for various photon energies separated by 20 meV.

3.2.7. *Franz-Keldysh oscillations*

An electric field produces a separation $\Delta E = eFD$ between single-well levels and eventually leads to the splitting of the miniband into a

Stark ladder. For very small fields and/or very strong coupling, when the separation is smaller than the broadening of the levels, the miniband picture is still valid. In this case the electric field tilts the miniband, the energy drops eFD per period.

In this regime the absorption spectrum presents oscillations above the bandgap that are different from Wannier-Stark oscillations. They are the Franz-Keldysh oscillations,[67,68] which have been extensively observed in bulk semiconductors.[69] In contrast to the Wannier-Stark oscillations, their period increases as $F^{2/3}$. They appear only when the superlattice states are very delocalized. At a finite electric field an energy level crosses the miniband edges at two points in the z-direction. Between these two points the envelope of the superlattice wavefunction has an oscillatory behavior. The regions outside these points are forbidden and in them the envelope decays exponentially. The oscillator strength of an optical transition, which is roughly proportional to the overlap of the electron and hole states, depends on the transition energy. For photon energies below the bandgap the absorption decays exponentially because the overlap is only between the exponential tails of the electron and the hole wavefunctions. Above the bandgap, however, the absorption spectrum exhibits oscillations because the overlap is between wavy sections of the envelope of the superlattice wavefunctions for the conduction and the valence bands. As for interfering waves, the intensity reaches maxima or minima when the crests or the valleys, respectively, of both functions are aligned.

Figure 3.14 shows gray-scale maps of the differential photocurrent spectra calculated and measured[28] for a 3.1 nm/0.3 nm-GaAs/AlAs superlattice. The Franz-Keldysh oscillations form two fans that emerge form the two critical points of the electron miniband,[28,32,33] M_0 and M_1. The branches of the fans E_m move with the field as[28]

$$E_m = E_\nu + (-1)^\nu \hbar \left(\frac{e^2}{2 |\mu_z^\nu| \hbar} \right)^{1/3} x_m F^{2/3} . \qquad (3.21)$$

where E_ν denotes the energies of the critical points M_ν, x_m the nodes of the superlattice wavefunctions, and μ_z^ν the reduced superlattice effective mass at the critical point M_ν. At very low electric fields, when the miniband is not split yet into a Stark ladder, the Franz-Keldysh oscillations dominate the spectrum. At higher fields the Stark ladder oscillations form a fine structure coexisting with the Franz-Keldysh oscillations.[28] At moderate

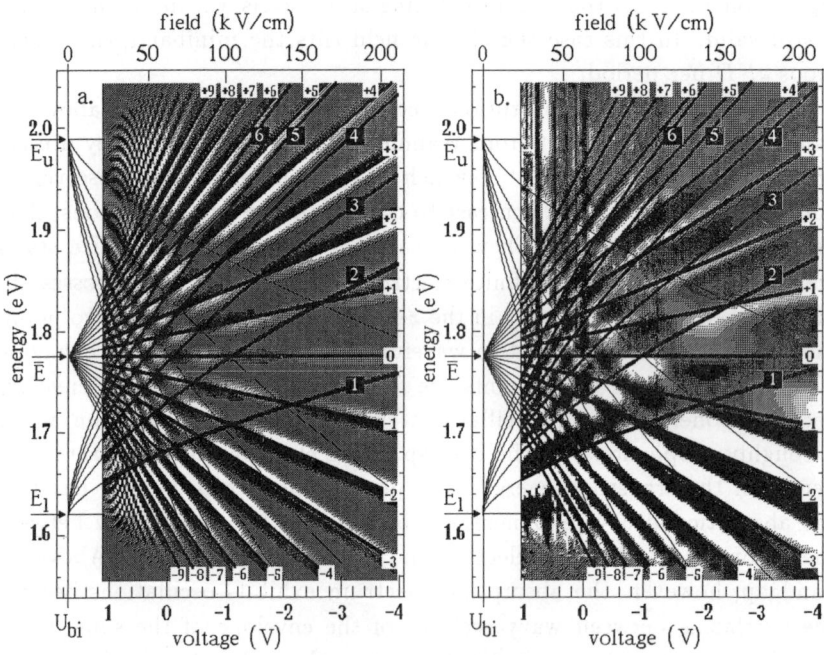

Fig. 3.14. Calculated (a) and measured (b) wavelength derivative of the photocurrent measured in a 3.1 nm/0.3 nm-GaAs/AlAs superlattice at 77 K as a function of photon energy and electric field. White points correspond to maxima. White labels give the indices p of Stark ladder transitions, whereas black labels stand for the Franz-Keldysh oscillations (Adapted from Ref. 28).

fields, when the Stark transitions are well resolved, still the wavy nature of the superlattice envelope function modulates their oscillator strength. This effect, which was discussed before, is reminiscent of the Franz-Keldysh oscillations already in the field range, where the miniband picture is not valid.

3.2.8. *Miscellaneous work*

We have discussed the more basic work on the Wannier-Stark localization. There is a great deal of work on closely related subjects which, because of the limited space, are only mentioned.

We have restricted ourselves to the case of type I superlattices where the electron and hole wells are in the same layers. The Stark localization has also been studied[23,70−72] for $\Gamma - X$ superlattices in which the lowest electron states are at the X minima of the barrier layer. In this case the optical transitions are indirect in real space and in k-space as they involve electrons and holes in different layers and at different points of the Brillouin zone.

The presence of two minibands induces at high fields a delocalization of states due to the resonance between the various Stark ladders.[25,52,73−75] The delocalization of superlattice states is also observed due to the resonance with states from enlarged wells[76−79] or the edge wells[80] in finite superlattices. In superlattices, where the gap increases linearly from well to well, minibands form at finite electric fields that are different for electrons and holes.[81] Additional localization effects are introduced by a magnetic field.[82−87] Finally, the formation of the Stark ladder has been used to induce doubly and triple Raman scattering when the LO phonon energy equals a multiple of the separation between Stark ladder levels.[26,27]

3.2.9. *Electro-optic applications*

The Stark localization in superlattices has already found application in a number of photonic devices. They are based on the electric-field induced changes in the optical properties or electro-optic effects[88] mainly near the absorption edge.

The most immediate application is the electro-optic modulator, in which an external voltage switches a light beam on and off. Usually the modulator is designed for a photon energy equal to the excitonic absorption edge of the superlattice. In this case the modulator is normally off, i.e., it is opaque for zero and moderate fields, but the blue shift of the absorption edge makes it transparent. This is in contrast with multiple-quantum-well modulators based on the quantum confined Stark effect.[88] Here the operating wavelength is slightly below the excitonic absorption edge near flat band and the device is normally on. The off state is produced at a finite field because of the red shift of the edge that makes the device opaque. In uncoupled quantum wells the shift is small and depends quadratically on the electric field, whereas in a superlattice the shift of the interwell transitions is linear with the field and is very large for large indices p. Fig. 3.15 shows a superlattice blue shift modulator in the perpendicular configuration. The

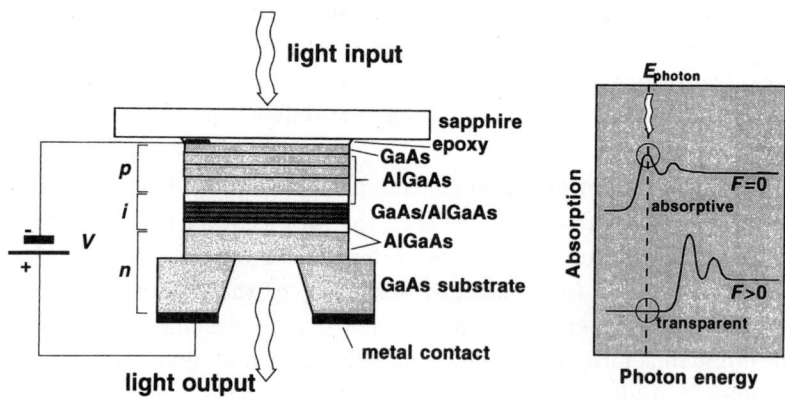

Fig. 3.15. Electro-optic modulator based on the electric-field-induced blue shift of the absorption edge in superlattices.

light beam passes through the device perpendicular to the layer planes. For GaAs/AlGaAs superlattices grown on GaAs substrates the substrate is opaque for the wavelengths of interest because the absorption edge of a quantum well is above that of the bulk material. Therefore, for modulators working in transmission the substrate has to be etched away in the central part.

In waveguide modulators (Fig. 3.16) the light propagates along the superlattice layers. The interaction length of the light with the superlattice is larger than in perpendicular modulators. In order to have it well confined, intermediate layers with a refractive index larger than the external layers are grown on both sides of the superlattice.[37,38] This structure is very similar to that of separate confinement heterostructure lasers (SCH) making it easy to monolithically integrate lasers and modulators for Q-switching.

In reflection modulators a highly reflective mirror is grown between the superlattice and the substrate. In this configuration the input and output light beams enter and exit the device through the top surface, respectively, and the absorptive substrate is avoided. The mirror is a Bragg reflector made by a stack of alternating quarter-wave n-doped layers with

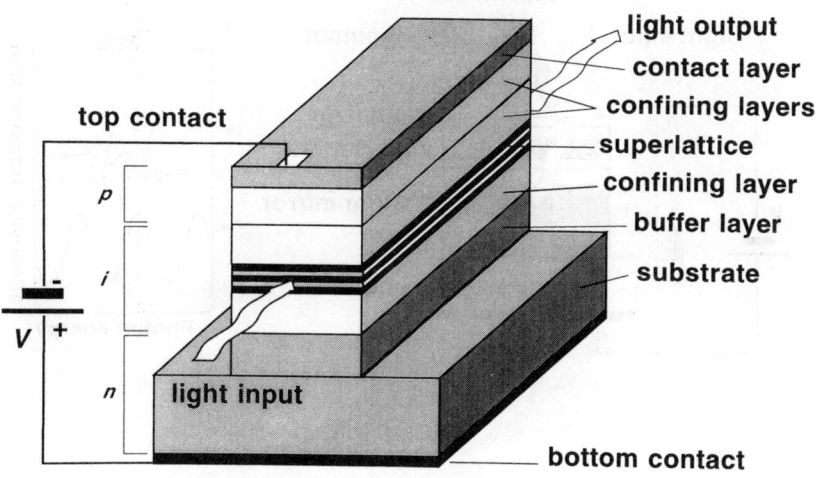

Fig. 3.16. Superlattice electro-optic modulator in a waveguide configuration.

different refractive indices such as GaAs and AlGaAs. These layers are grown as the superlattice and are carefully designed to have a reflectivity maximum at the operating wavelength, i.e., at the excitonic absorption edge of the superlattice at zero field. In these devices, both transmission and reflection modulators, the modulation is very modest because it is limited by the weak superlattice absorption. The contrast ratio between the reflectance in the on and off states is less than 5:1.

For real applications higher contrast ratios are needed. With super-lattice asymmetric Fabry-Perot reflection modulators[36] (Fig. 3.17) contrast ratios of more than 100:1 for the on:off states have been achieved.[39] The bottom mirror of the Fabry-Perot interferometer is a Bragg reflector of quarter-wave alternating layers. The top mirror is simply the semiconductor-air interface, which has a low reflectivity. One of the resonances or reflection energies of the interferometer coincides with the excitonic absorption edge of the superlattice. The light beams reflected at the top and at the bottom mirrors have opposite phases. The absorption of the superlattice at zero bias (off state) is such that the amplitudes of both beams are the same.

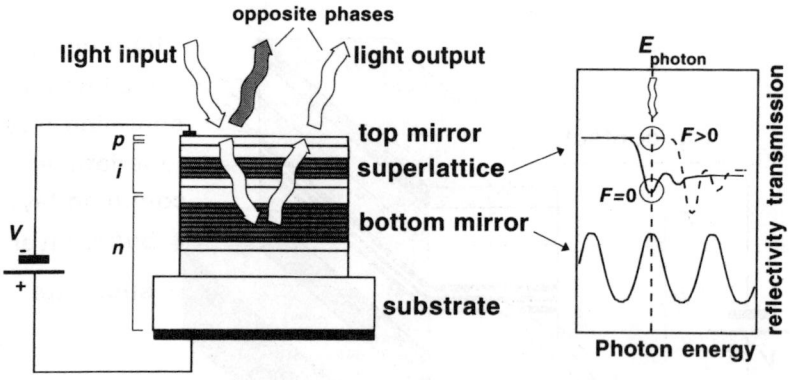

Fig. 3.17. Superlattice asymmetric Fabry-Perot electro-optic modulator.

Therefore, there is destructive interference, and no light is reflected from the device. For a finite bias (on state) the blue shift makes the superlattice transparent, and the bottom mirror reflection dominates giving a large overall reflection. To fulfill all these conditions the device must be carefully designed and constructed.

Another electro-optic device is the SEED (for Self-Electro-optic Effect Device).[88] The first SEED's were based on the quantum confined Stark effect in uncoupled quantum wells. However, the blue shift and the rich negative differential resistance in superlattices give better results. SEEDs based on the Stark localization have been recently made.[34,35,39,89] The simplest superlattice SEED consists of a p-i-n superlattice modulator in series with a resistor (Fig. 3.18). The circuit is biased with a constant voltage. A light beam of power P_{in} passes through the modulator and exits with a smaller power P_{out} due to the absorption in the modulator.

What is interesting about the SEED is the dependence of P_{out} on P_{in} (Fig. 3.18). Under the right conditions it shows bistability with two possible values of P_{out} depending on whether P_{in} is increasing or decreasing. The SEED is an electro-optic device because it is based on an electrooptic effect. Considered as a whole, e.g., the p-i-n diode and the external electric circuit,

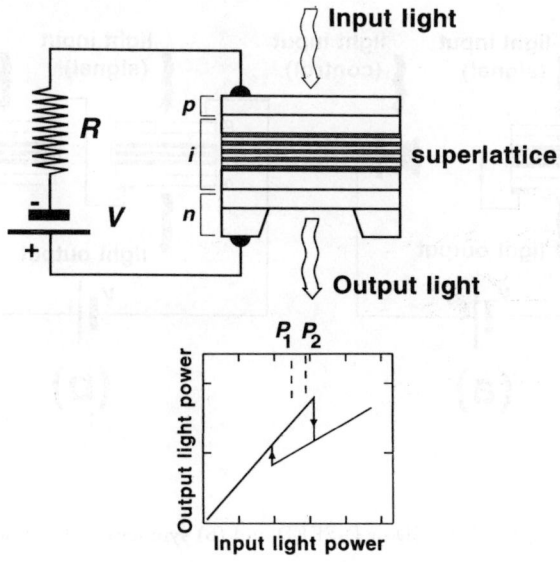

Fig. 3.18. Self-Electro-optic Effect Device (SEED) based on a blue shift superlattice modulator.

however, it is also a nonlinear optical device with promising applications in optical computing.

The simple SEED can be improved by replacing the resistor with a photodiode (diode SEED or D-SEED) or another SEED (symmetric SEED or S-SEED) (Fig. 3.19). D-SEED[34,39] and S-SEED[40] based on the Stark localization have been fabricated. With superlattices it has been possible to achieve multistabilities.[35,40]

SEEDs have been reported for GaAs/AlGaAs[34,39] and for GaAs/AlAs superlattices.[35,40,89] The influence of the miniband width on the device performance has been analyzed.[89] The SEED has been used as a building block for some photonic devices where light is controlled by light. These kind of devices are essential for the development of optical computers. Although the bistability of the SEED makes it very appropriate for digital signal processing, an optical equivalent of the electronic transistor, which is an analog device, has also been realized with a superlattice S-SEED.[41]

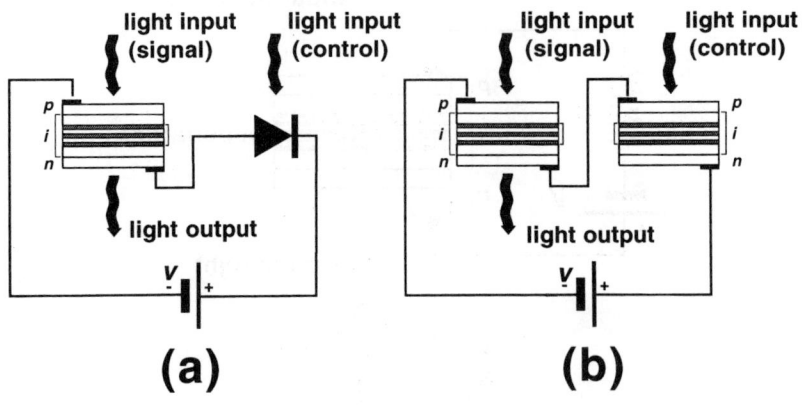

Fig. 3.19. (a) Diode SEED or D-SEED and (b) symmetric SEED or S-SEED.

3.3. Bloch oscillations

We now focus on the dynamic aspects of semiconductor superlattices in the presence of electric fields. In section 3.3.1 we discuss the principal similarities of Bloch oscillations and Wannier-Stark ladder transitions. We point out that the conditions for the observability of Bloch oscilations and Wannier-Stark ladders are the same. We then dicuss in section 3.3.2, why four-wave mixing experiments are a suitable technique to observe Bloch oscillations in the time domain. In section 3.3.3 and section 3.3.4 we present results on semiconductor superlattices with narrow and wide minibands and show that LO-phonon scattering in wide minibands controls the field-onset of both, Bloch oscillations and Wannier-Stark ladders. In addition, Coulomb effects are discussed in sections 3.3.5 and 3.3.6. At low electric fields, field-ionization of excitons is clearly observed in the time domain for superlattices with wide minibands. A more subtle problem is how many-body Coulomb effects, which generally influence the temporal dynamics of four-wave mixing experiments, are changed by the applied electric field.

3.3.1. *Introduction*

Electrons in a band of a periodic crystal are described by Bloch eigenfunctions

$$\psi(\vec{k}, \vec{r}, t) = u(\vec{k}, \vec{r}) \, exp \left[i\vec{k} \cdot \vec{r} - i \int^t \frac{E(\vec{k})}{\hbar} dt' \right] , \qquad (3.22)$$

where $E(\vec{k})$ denotes the band dispersion. In the presence of an electric field $\vec{F}$, the equation of motion for an electronic Bloch state is given by the so-called acceleration theorem[90-94]

$$\hbar \, \dot{\vec{k}} = e \, \vec{F} . \qquad (3.23)$$

A Bloch electron composed of Bloch states moves through $\vec{k}$-space with a field-dependent velocity as depicted in the lower part of Fig. 3.20. Due to

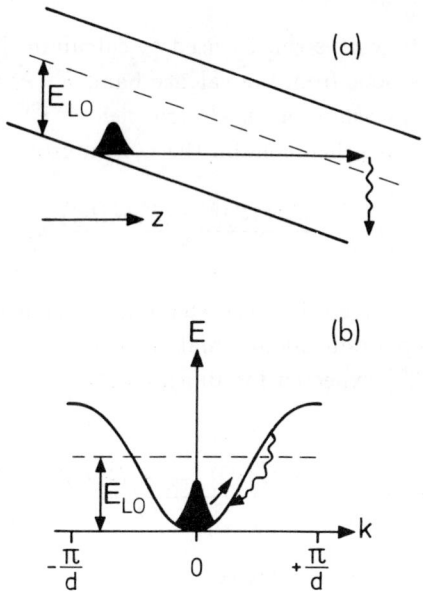

Fig. 3.20. Schematic illustration for the field-induced coherent motion of an electronic wavepacket initially created at the bottom of a miniband having a width exceeding the LO-phonon energy E_{LO}. Scattering by LO-phonons is illustrated in the real space picture (upper part) and the k-space picture (lower part).

Bragg-reflections at the Brillouin-zone (BZ) boundary the electron executes so-called Bloch oscillations within a single band with a time period

$$T_B = \frac{h}{e\,F\,D}, \tag{3.24}$$

where D is the lattice constant. To a good approximation, the electronic wavefunctions in the presence of an electric field are still given by Eq. 3.23, if $\vec{k}$ in Eq. 3.23 is made time-dependent according to the acceleration theorem given by Eq. 3.24. However, these so-called Houston-functions[95] neglect the 'leakage' of electrons from one band into another and thus the possibility of field-induced electron tunneling into higher bands (Zener tunneling).[96] In addition, in real crystals electron scattering due to phonons, impurities, etc. limit the coherent motion described by the acceleration theorem to a finite time τ_{sc}.[97] We note that the acceleration theorem is quantum mechanically correct even in the presence of Zener tunneling, whereas the concept of Bloch oscillations within a single band is only valid for times less than the Zener tunneling time.[98]

The Houston-functions can be used to calculate the matrix-elements for the optical transitions from the valence-band $E_v(\vec{k})$ to the conduction-band $E_c(\vec{k})$ in the presence of an electric field.[99,100] The resulting one-particle matrix-elements then contain the non-sinusoidal phase-factor[99]

$$exp\left[i \int^t \frac{E_c(\vec{k}(t')) - E_v(\vec{k}(t'))}{\hbar}\,dt'\right]. \tag{3.25}$$

Due to the periodic traversal of the electron through the BZ the phase-factor in Eq. 3.26 is periodically modulated with a time-period equal to the period $T_B = h/(eFD)$ expected for Bloch oscillations. The corresponding Fourier component

$$\frac{1}{2\pi} \int_{-\infty}^{+\infty} dt\, exp\left[i \int^t \frac{E_c(\vec{k}(t')) - E_v(\vec{k}(t')) - \hbar\,\omega}{\hbar}\,dt'\right] \tag{3.26}$$

then exhibits sidebands with energies

$$E_n = E_0 + n\,e\,F\,D\,, \qquad n = 0, \pm 1, \pm 2, \ldots \tag{3.27}$$

giving rise to the Wannier-Stark (WS) ladder transitions as depicted in the band scheme of Fig. 3.21.

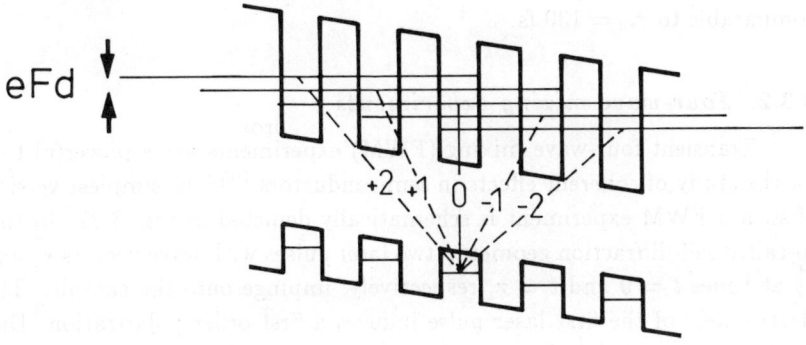

Fig. 3.21. Bandscheme of the conduction- and valence-band potential profiles of a superlattice in the presence of an electric field F.

The requirement for both, the formation of the WS ladder and the occurrence of Bloch oscillations, is the same, namely that the period T_B for Bloch oscillations is less than the scattering time τ_{sc}.[97,101] The electron must be able to make at least one oscillation through the BZ before being deflected. In the case that for all reasonable electric fields the scattering time τ_{sc} is shorter than the Bloch oscillation period T_B, no WS transitions but Franz-Keldysh oscillations (FKO) can be observed in the absorption spectrum (see section 3.2.7).[102,103] For the calculation of FKO only $E(\vec{k})$-states next to $\vec{k} = 0$, where the parabolic band approximation applies, are considered in Eq. 3.26 and 3.27,[99,100] since electrons are assumed not to reach large $\vec{k}$-states due to the strong scattering. The transition from the *band regime* to the *WS regime* occurs at an applied electric field where electrons are fast enough in $\vec{k}$-space to beat the scattering and to reach the BZ boundary coherently.[101]

In particular, electron scattering by optical phonons is expected to efficiently damp Bloch oscillations and broaden the homogeneous linewidth of WS transitions, since electron-LO-phonon scattering times are less than 130 fs in the GaAs/AlGaAs system.[104] Semiconductor superlattices offer the unique opportunity to tailor the lattice constant D and the miniband width Δ. As depicted in Fig. 3.20, we expect an efficient damping of Bloch oscillations when the miniband width Δ exceeds the LO-phonon energy.

Then the transition from the *(mini)band regime* to the *WS regime* occurs at high electric fields, when the time-period T_B for Bloch oscillations becomes comparable to $\tau_{sc} = 130$ fs.

3.3.2. *Four-wave mixing experiments*

Transient four-wave mixing (FWM) experiments are a powerful tool for the study of coherent effects in semiconductors.[105] The simplest version of such a FWM experiment is schematically depicted in Fig. 3.22. In this so-called self-diffraction geometry, two laser pulses with wave vectors $\vec{k}_1$ and $\vec{k}_2$ at times $t = 0$ and $t = \tau$, respectively, impinge onto the sample. The electric field of the first laser pulse induces a first order polarization. Due to the nonlinear optical interaction, the electric field of the second laser pulse then creates a third-order polarization, which is the source for the diffracted FWM signal in the phase-matched direction $2\vec{k}_2 - \vec{k}_1$. The FWM signal can then be measured *time-integrated* as a function of the time-delay τ between the two laser pulses by using a slow photodetector. In addition, it is possible to time-resolve the FWM signal for each time-delay τ by using an up-conversion technique. In such *time-resolved* FWM experiments photon echos are observed in case of inhomogeneously broadened two-level systems.[106]

In 1973, Zakharov and Manykin[107] theoretically analyzed the effect of a constant electric field on the diffracted photon echo in a transient FWM experiment for a bulk semiconductor. These authors found that the photon echo signal should reflect the temporal development of the wavevector $\vec{k}(t)$. However, due to the small lattice constant in bulk crystals and due to efficient carrier scattering, Bloch oscillations or WS ladders have not been observed in bulk semiconductors. In a more specific theoretical treatment, von Plessen and Thomas[108] proposed to detect Bloch oscillations directly in the time domain by performing FWM experiments on an electrically biased superlattice. Here, we summarize their main results. In k-space, the ensemble of vertical optical transitions ($\Delta k = 0$) between the valence-miniband and the conduction-miniband states can be viewed as an inhomogeneously broadened ensemble of two-level systems with transition frequencies $\omega(k)$. When a spectrally broad (short) laser pulse excites all of these two-level systems at $t = 0$, their dipole moments are initially in phase leading to a macroscopic (first order) polarization. Even without an applied electric field, the phases $\omega(k)t$ of the individual dipole moments evolve differently

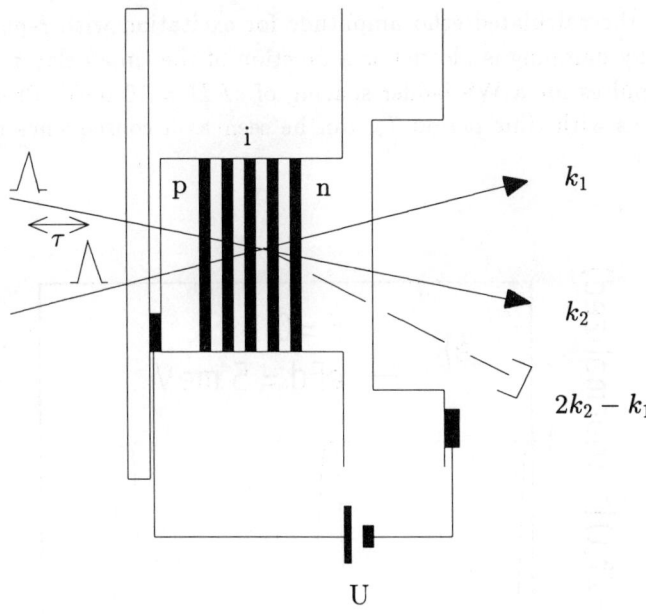

Fig. 3.22. Scheme of a time-intagrated FWM experiment performed on an electrically biased superlattice.

in time due to the different (but fixed) phase velocities $\omega(k)$. As a consequence, the macroscopic polarization vanishes. However, for uncoupled two-level systems the second pulse at $t = \tau$ leads to a zero phase shift at $t = 2\tau$. The recovered macroscopic (third order) polarization then leads to the emission of a photon echo.[109] With an applied electric field, the temporal variation of the quasi-momentum $k(t)$ (Eq. 3.24) implies that the phases $\int \omega(k(t))dt$ of the dipole moments now evolve according to the temporally varying phase velocities $\omega(k(t))$. As a result, the nonlinear interaction due to the second laser pulse generally does not lead to a perfect zero phase shift at the instant the photon echo should arise ($t = 2\tau$). Only for specific values of the field, such that the electron is able to traverse the entire BZ and return to the initial state during the time between the pulses, i.e., for $t = T_B$, the radiators will again be in phase at $t = 2\tau$. Altogether, the transient FWM signal for a biased superlattice should exhibit periodic modulations as a function of the time interval τ between the pulses. In Fig. 3.23, the calculated echo amplitude for excitation with δ-pulses and without any damping is plotted as a function of the time delay τ between the laser pulses for a WS ladder spacing of $eFD = 10$ meV. Pronounced modulations with time period T_B can be seen as a consequence of Bloch oscillations.

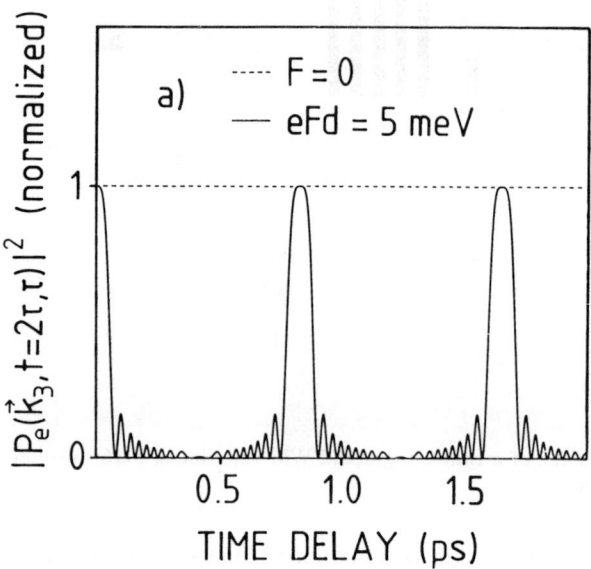

Fig. 3.23. Calculated FWM signal for a WS ladder spacing $eFd = 10$ meV as a function of the time delay τ between the δ-function laser pulses and in the absence of damping.

3.3.3. *Bloch oscillations in narrow minibands*

We now review and discuss results of sub-picosecond FWM experiments performed on a $GaAs/Al_xGa_{1-x}As$ superlattice sample. The intrinsic superlattice structure consists of 91 periods of 9.5 nm GaAs and 1.5 nm $Al_{0.3}Ga_{0.7}As$ embedded in a p-i-n diode.[110] The conduction miniband width of this superlattice sample is approximately 21 meV and is thus less than the LO-phonon energy. Fox et al.[59] studied the linear optical properties of this superlattice sample in detail. Fig. 3.24 shows low-temperature photocurrent spectra for electric fields up to 2.5×10^4 Vcm^{-1} (-1 V reverse bias). A well-defined heavy-hole Stark-ladder, i.e., a fanlike behavior of the state energies as a function of field according to Eq. 3.28 is evident. The *WS regime* already starts at fields lower than 4×10^3 Vcm^{-1}, where WS ladder transitions with eFD spacings of less than 4 meV are still resolved in the photocurrent spectrum. We argue that the absence of LO-phonon scattering within the 21 meV wide miniband prolongs τ_{sc} and allows electrons to perform 'slow' Bloch oscillations at low electric fields.

In Fig. 3.25, the diffracted FWM signal is shown as a function of the time delay τ between the 110 fs laser pulses for a WS ladder spacing of 6.2 meV. A pronounced beating with a time period of 0.7 ps is observed in good agreement with the time period $T_B = h/(eFD) = 0.67$ ps expected for Bloch oscillations. By lowering the electric field T_B-values as long as

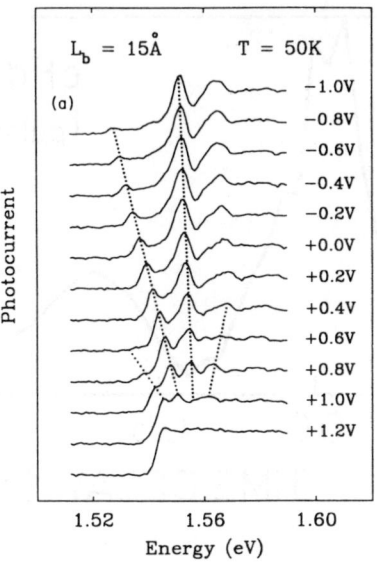

Fig. 3.24. Photocurrent spectra for a superlattice with a conduction-miniband width of about 21 meV (taken from Ref. 58).

1.4 ps corresponding to $eFD = 3$ meV were observed.[110] For a very similar superlattice structure, also with a miniband width less than the LO phonon energy, T_B-values as long as 1.7 ps corresponding to WS ladder spacings as low as $eFD = 2.4$ meV were measured by Leo et al.[111] We conclude that the scattering times τ_{sc} for electrons and holes must be on a picosecond time scale for all k-states throughout the mini-BZ. Otherwise, electrons with such low velocities could not coherently move in k-space for such long time intervals. Probably, electron-acoustic phonon scattering is the limiting scattering mechanism for these narrow miniband superlattices.

In a theoretical paper, Bouchard and Luban[112] showed that dipole radiation in the terahertz range should be detectable as a consequence of optically generated Bloch oscillations as initially proposed by Esaki and Tsu.[113] Waschke et al.[114] have indeed detected coherent electromagnetic radiation originating from Bloch oscillations in a superlattice structure with a narrow miniband width.

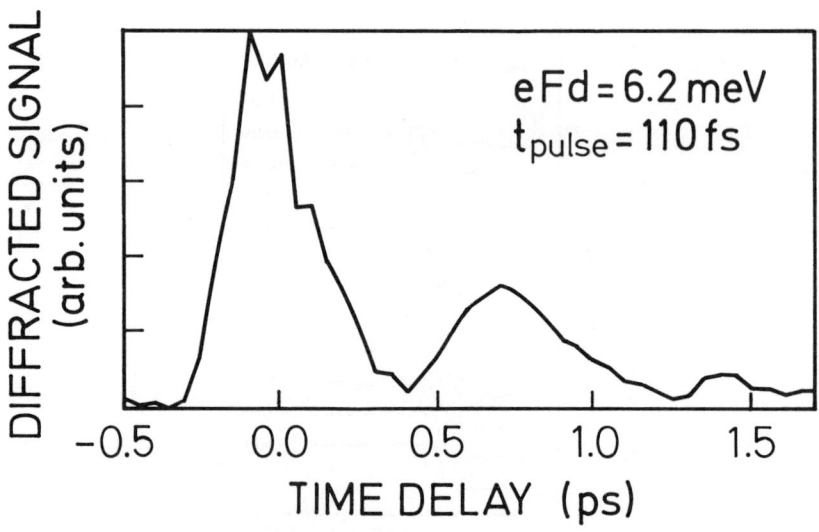

Fig. 3.25. FWM signal versus time delay for $eFD = 6.2$ meV using 110 fs laser pulses.

3.3.4. *LO-phonon scattering in wide minibands*

As already discussed above electron scattering by LO-phonons has to be considered when the miniband width of the superlattice exceeds the LO-phonon energy. In the presence of an electric field, electrons move in the $E(k)$-superlattice dispersion as depicted in the lower part of Fig. 3.20. When they reach a kinetic energy higher than the LO-phonon energy they might be scattered by emission of a LO-phonon. Whether or not the electron performs Bloch oscillations by undergoing Bragg reflections at the Brillouin-zone boundary now depends on the electron's k-space velocity. In an equivalent real space scheme as shown in the upper part of Fig. 3.20, a standing wave within the miniband is only formed when the electron coherently reaches the upper edge of the miniband and becomes reflected. Again, whether or not a standing wave (WS wavefunction) is formed depends on the field-induced velocity of the electron.

Any loss of coherence due to scattering processes will, in the time domain, damp the third-order polarization. The damping corresponds in the frequency domain to a homogeneous broadening of the WSL transitions. The WSL cannot be resolved optically unless the energy separation between subsequent WSL levels is larger than the homogeneous linewidth, Γ_{hom}. In other words, for the WSL to be resolved, the applied electric field must be larger than some critical field F_c, which is given by

$$e\,F_c\,D \;=\; \Gamma_{hom}\;. \qquad (3.28)$$

The homogeneous linewidth is related to the scattering time τ through $\Gamma_{hom} = 2\hbar/\tau$. The critical field is thus given by $eF_cD = 2\hbar/\tau$; apart from a factor of 2, this value agrees with the usual criterion for WS quantization.[101] We note that in real superlattice samples inhomogeneous broadening of the optical transitions cannot be avoided. Consequently, when analyzing linear optical spectra of WSL, Γ_{hom} in Eq. 3.29 has to be replaced by the sum of the homogeneous linewidth and the inhomogeneous broadening, $\Gamma_{hom} + \Gamma_{inhom}$. This complication is not present in transient four–wave mixing experiments since FWM, unlike linear spectroscopy, gives direct access to the homogeneous linewidth.[115]

Previous estimates of the critical field used a single scattering time for the entire miniband. However, this assumption represents a serious oversimplification. In bulk semiconductors at $F = 0$, energy conservation requires that electrons in conduction band states separated from the band minimum by less than one LO phonon energy cannot become involved in the emission of an LO phonon. Therefore, scattering via LO-phonon emission, which is the dominant scattering mechanism at low temperatures, does not take place below this threshold energy. A similar effect is to be expected for semiconductor superlattices. In analogy to the bulk case, intra-miniband scattering of electrons via LO-phonon emission will not occur below the corresponding threshold energy, E_{LO}. For GaAs/AlGaAs superlattices, E_{LO} is given by the energy of the GaAs-like LO-phonon mode, and thus lies at 36 meV above the bottom of the miniband.[27] This role of E_{LO} will be of importance also for $F \neq 0$. We conclude that a miniband electron performing a field-induced motion according to the acceleration theorem will experience a reduced scattering rate in regions of the mini-Brillouin-zone, where $E(k(t)) < E_{LO}$, and the full scattering rate, where $E(k(t)) > E_{LO}$.[116]

This effect is important for the critical field F_c in the following sense. In general, the total scattering time will lie in between $\tau_<$ and $\tau_>$, where $\tau_<$ and $\tau_>$ are the scattering times below and above E_{LO}, respectively. Its precise value will depend on the width of the electron miniband Δ_e. In particular, an interesting situation arises when $\Delta_e < E_{LO}$. Then scattering due to LO-phonon emission is completely excluded. We therefore expect eF_cD to be much smaller in superlattices with narrow minibands than in those with wide minibands.

We study a superlattice sample with miniband width greater than $E_{LO} = 36$ meV. The sample was grown by molecular-beam epitaxy on a n-doped GaAs substrate. The superlattice (SL) structure is located in the intrinsic region of a p-i-n diode and consists of 100 periods of 3.0 nm GaAs and 3.0 nm $Al_{0.3}Ga_{0.7}As$. Kronig-Penney calculations yield $\Delta_e = 62$ meV for the electronic miniband width and $\Delta_{hh} = 5$ meV for the heavy-hole miniband width.

To determine the critical field for the onset of the WSL and the Bloch oscillations, we perform photocurrent[14] and transient four-wave mixing (FWM) experiments.[117] Fig. 3.26 shows the results of the photocurrent measurements taken at 77 K. The energetic positions of the photocurrent peaks are plotted versus the nominal field determined from the voltage applied to the p-i-n structure. We estimate that the actual field experienced by the superlattice is smaller than the nominal field by approximately 6 kVcm^{-1}, which could be due to screening of the field by the photogener-

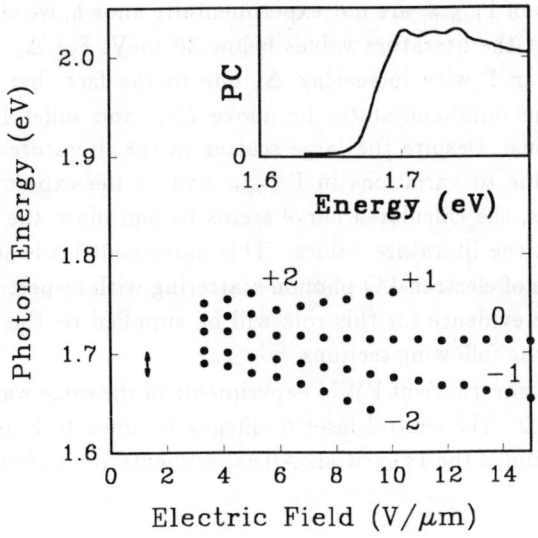

Fig. 3.26. Photocurrent peak positions versus applied electric field. Shown in the inset is a typical photocurrent spectrum for an applied field of $F = 43$ kVcm^{-1}.

ated carriers.[117] The plot clearly shows the fan of the heavy-hole WSL transitions. The photocurrent peaks completely disappear below a field corresponding to a minimum WSL spacing of $eF_cD \approx 14$ meV. For comparison, Fig. 3.27 shows the eF_cD values which can be extracted from WSL fan charts measured by other groups on various GaAs/Al$_x$Ga$_{1-x}$As superlattices ($x = 0.3 \ldots 1$). It can be seen that the minimum energy spacings obtained for wide minibands[14,16,27,29,118,119] all lie above 7 meV, whereas those with narrow minibands[59,70,111,117] lie around 5 meV. As will be shown in the following model calculation, this finding can be explained by the suggested role of electron-LO phonon scattering in wide minibands.

Our calculation is based on a simple one-dimensional tight-binding model of a superlattice. We consider a cosine miniband dispersion and include the electric field via the acceleration theorem.[108] We assume $\tau_> = 132$ fs[104] and $\tau_< = 1.5$ ps for the scattering times above and below E_{LO}, respectively. We calculate the total linewidth of the WSL transitions, $\Gamma = \Gamma_{hom} + \Gamma_{inhom}$, which is expected to agree with eF_cD according to Eq. 3.29.

The result of the calculation is shown as a solid line in Fig. 3.27; since the precise values of Γ_{inhom} are not experimentally known, we chose $\Gamma_{inhom} = 4.3$ meV to fit the literature values below 36 meV. For $\Delta_e > 36$ meV we obtain a rise in Γ with increasing Δ_e due to the fact that an increasing fraction of the miniband states lie above E_{LO} and suffer fast scattering by LO phonons. Despite the large scatter in the literature values, which is probably due to variations in Γ_{inhom} and in the experimental signal-to-noise ratios, the calculated curve seems to reproduce the general trend manifested in the literature values. This agreement lends support to the suggested role of electron-LO phonon scattering with respect to the critical field. Further evidence for this role will be supplied by the FWM results presented in the following sections.

We perform transient FWM experiments in the same way as described in section 3.3.2. The central laser frequency is tuned to 3 meV below the spectral position of the 1s exciton. All experiments are performed at crystal

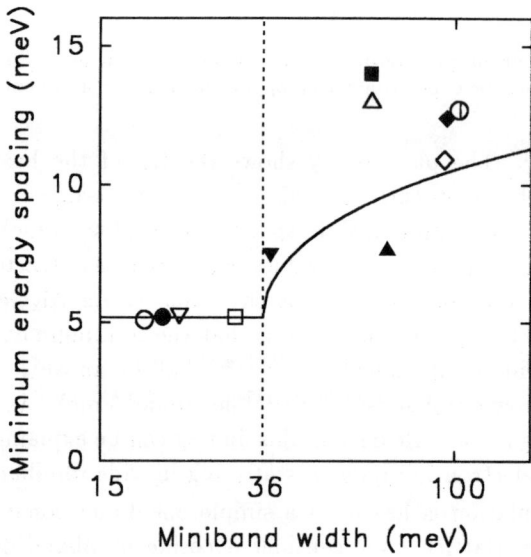

Fig. 3.27. Minimum energy spacings determined from Wannier-Stark ladder fan charts in the literature. Also shown is the calculated total linewidth Γ as a function of the miniband width.

temperatures below 10 K. Fig. 3.28a and Fig. 3.28b show, as solid lines, FWM transients measured at 0 and 28 kVcm^{-1} nominal field, respectively. Again, as in the case of the photocurrent measurements, we have to keep in mind that the actual field experienced by the superlattice is slightly smaller than the nominal field calculated from the applied voltage. Fig. 3.28b shows that at 28 kVcm^{-1} nominal field, which corresponds to $eFD = 13.2$ meV and $T_B = 0.35$ ps, the decay time of the FWM signal is drastically reduced in comparison with the flat-band case, but no periodic modulation indicative of Bloch oscillations is observable. This is in striking contrast to the transient FWM experiments performed on superlattices with narrow minibands as discussed in the previous section. As will be shown in the following theoretical analysis, this discrepancy can only be explained by taking into account electron-LO phonon scattering in a wide miniband.

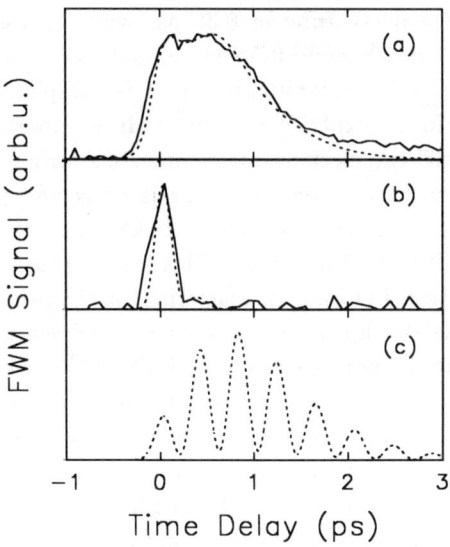

Fig. 3.28. Experimentally (solid line) and theoretically (dashed line) determined FWM signals for applied fields $F = 0$ (a) and $F = 28$ kVcm^{-1} (b). (c) shows the calculated four-wave mixing signal for $F = 28$ kVcm^{-1} with the electron–LO phonon interaction neglected.

Our analysis is based on the same superlattice model as above. Additionally, we include excitonic effects through a δ-like contact potential,[120] i.e., we consider the Coulomb interaction within the same well of the superlattice. For this model, we numerically solve the semiconductor Bloch equations[121] in third order in the optical fields and obtain the diffracted FWM signal from the solution. The parameters used in the calculations are $\Delta_e = 62$ meV for the electronic miniband width, $\Delta_{hh} = 5$ meV for the heavy-hole miniband width, 4.5 meV for the exciton binding energy,[64] and $\tau_> = 132$ fs[104] and $\tau_< = 1.5$ ps for the scattering times above and below E_{LO}, respectively. We use the same pulse duration and spectral detuning as in the experiment and Γ_{inhom} as a fit parameter.

Fig. 3.28a and Fig. 3.28b show, as dashed lines, the calculated FWM signal transients for $F = 0$ and $F = 28$ kVcm^{-1}, respectively. From fitting the $F = 0$ curve, we obtain an inhomogeneous broadening of $\Gamma_{inhom} = 3.5$ meV, which is a reasonable value. The transient for $F = 28$ kVcm^{-1} nominal field is calculated using the same value for Γ_{inhom}. It exhibits a second peak at a delay time of 0.35 ps, which is associated with the completion of one Bloch oscillation cycle, but is so strongly damped in comparison with the first maximum that it is completely masked by the background noise in the experimental curve. It is important to realize that when 'switching off' the electron-LO phonon scattering in the calculation by setting $\tau_>$ equal to $\tau_<$, then damping is weak and pronounced Bloch oscillations appear in the calculated $F = 28$ kVcm^{-1} transient (Fig. 3.28c) in disagreement with the experiment. This leads to the obvious conclusion that it is the electron-LO phonon scattering mechanism which suppresses all evidence for the Bloch oscillations in the experiment.

We note that the nominal field of 28 kVcm^{-1} corresponds to a WSL spacing of 13.2 meV. This value is close to our experimentally determined value of $eF_cD = 14$ meV for the formation of the WSL. It is natural to assume that one would only have to increase the applied field beyond F_c to overcome the electron-LO phonon scattering and observe Bloch oscillations. However, this would imply a Bloch oscillation time T_B too close to our present time resolution, which is limited by the pulse duration and amounts to a theoretical value of approximately 0.25 ps for the detection of Bloch oscillations. Despite this limitation, we are able to draw the important conclusion that the Bloch oscillations are damped by fast scattering in a field range where they have been shown to be observable in superlattices with narrow minibands.[110,111] In addition to the photocurrent measurements, the FWM results thus lend further support to our suggestion concerning the role of electron-LO phonon scattering in controlling the critical field F_c.

3.3.5. *Field-induced exciton ionization in a wide superlattice miniband*

Recently, it has been shown for the case of unbiased quantum well structures that the simultaneous optical excitation of the bound exciton (1s, 2s, ...) and unbound electron-hole continuum states leads to a characteristic quantum beat behavior of the time-integrated FWM transient.[122] As a consequence of a destructive interference effect between the 1s exciton transition and the continuum transitions, a pronounced initial decrease of the FWM signal is observed. In addition, a moderate $1s - 2s$-exciton beating occurs for larger time delays. Altogether, such FWM experiments allow the experimental observation of the coherent dynamics of exciton wave packets. A natural extension of these experiments is the investigation of exciton wave packets in the presence of an electrical field.

Very recently, von Plessen et al.[123,124] have performed time-integrated FWM experiments using 55 fs laser pulses on the superlattice sample with wide minibands, which has been described in detail in section 3.3.4. The central photon energy is tuned 6 meV below the band-gap exciton resonance. In the semi-logarithmic plot of Fig. 3.29, time-integrated FWM transients for three distinct applied voltages are shown. The applied voltage of -1.5 V (forward bias) corresponds to flat-band condition, i.e., the expected electric field in the intrinsic superlattice region is expected to vanish. A beating of the FWM transient is observed, which shows the characteristic initial feature of an excitonic wave packet. At -0.5 V (forward bias) the beating is still present but less pronounced. However, at 0 V a purely exponential decay of the FWM signal is observed. For even higher electric fields (not shown in Fig. 3.29) a continuous shortening of the FWM decay is observed.

In a simplified picture, the suppression of the 1s-continuum beating (initial pronounced signal decay and recovery) with increasing electric field can be understood in the following way.[123,124] In the presence of an electric field the tunneling rate of an electron-hole pair out of the 1s exciton state into the electron-hole continuum states is limited by the height of the Coulomb wall. Accordingly, an increasing electric field should lead to shorter tunneling times. In contrast, electron-hole pairs in the continuum states are free to move according to the acceleration theorem. Due to the efficient spatial separation of electron-hole pairs in the continuum, the contribution of continuum transitions to the FWM signal in Fig. 3.29 is more and more reduced with increasing electric field. Conclusively, the unmodulated and exponentially decaying FWM signal at 0 V in Fig. 3.29 solely originates from the $1s$ exciton resonance. The field-induced tunneling of the electron-hole pair out of the $1s$ exciton state, i.e., out of the tilted Coulomb-potential, is then responsible for the field-induced shortening of the FWM decay.

A quantitative comparison of the experimental FWM data with FWM calculations taking into account Coulomb effects and phonon-scattering allows the determination of the $1s$-exciton ionization times as a function of the applied electric field.[124] In Fig. 3.30, the ionization times are plotted

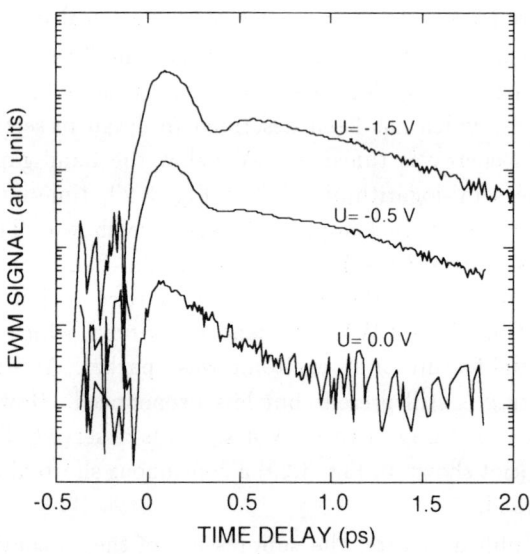

Fig. 3.29. Time-integrated FWM transients for the electrically biased GaAs/AlGaAs superlattice with wide minibands for three distinct voltages (forward bias) using 55 fs laser pulses. The curves are shifted in height to avoid overlapping.

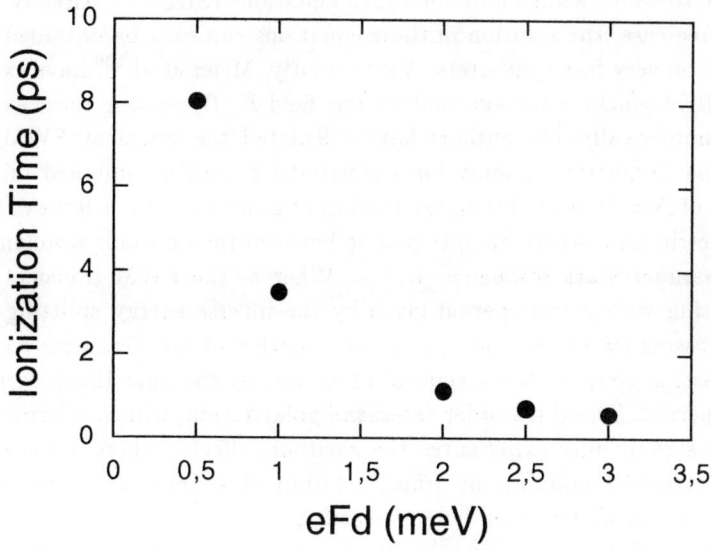

Fig. 3.30. Experimentally determined 1s-exciton ionization times versus the applied electric field. The reduction of the electric field by laser-induced screening has been taken into account in the analysis.

versus the electric field. For eFD-values exceeding ≈ 2 meV, the exciton ionization times lie in the subpicosecond regime.

3.3.6. *Coulomb effects*

As already discussed in section 3.2.6, the linear optical spectra of narrow-miniband superlattices, where the miniband width is of the order of the exciton binding energy, show a characteristic anticrossing behavior between the excitonic resonance and Wannier-Stark resonances.[60,62] It is now well known that the attractive Coulomb interaction between electrons and holes is responsible for the deviation from the simple single-particle behavior of the interband absorption.

During recent years, it has been shown that many aspects of time-resolved FWM experiments on semiconductors (without an applied electric field) can only be understood when many-body Coulomb effects are pro-

perly taken into account.[125−129] Calculations of FWM transients are based
on the so-called semiconductor Bloch equations (SBE).[121] Already for the
field-free case, the solution of these equations can only be obtained numerically on very fast computers. Very recently, Meier et al.[130] have extended
the SBE to include the external electric field F. By solving these extended
SBE numerically, the authors have calculated the transient FWM signal
and the Terahertz emission for a superlattice with a combined miniband
width of $\Delta = 10$ meV, for an exciton binding energy of $E_b = 9$ meV, and for
an electric field, where an anticrossing between the excitonic resonance and
the Wannier-Stark resonance occurs. Whereas the FWM transient shows
a beating with a time-period given by the inverse energy splitting at the
anticrossing ($\neq eFD$), the dynamical evolution of the Terahertz signal, in
contrast, is given by the inverse of eFD, i.e., by the pure Bloch oscillation
time-period. The third-order *interband* polarization, which determines the
FWM signal, fully experiences the excitonic effects, whereas the second-
order *intraband* populations, which determine the Terahertz signal, are not
influenced by the electron-hole interaction.

Von Plessen et al.[123,124] have recently shown that the FWM signal
strengths of superlattices with narrow and wide minibands show characteristic dependences on the applied electric field. Calculated FWM signal
strengths obtained by numerically solving the extended SBE and taking
into account LO-phonon scattering in wide minibands quantitatively agree
with the experimental results. This shows that both, many-body Coulomb
effects and phonon scattering in narrow and wide minibands, are important
ingredients for a complete theoretical description of time-resolved nonlinear
optical experiments on electrically biased superlattices.

3.4. Acknowledgments

We would like to thank many colleagues that have worked with us
on the various subjects throughout the years. Fernando Agulló-Rueda is
deeply in debt with E. E. Mendez, J. A. Brum, J. M. Hong, H. Ohno, L.
L. Chang, L. Esaki, H. T. Grahn, K. Ploog, and K. von Klitzing. Jochen
Feldmann would like to thank G. von Plessen and T. Meier for the innu-
merable contributions and J. Cunningham, E. O. Göbel, K. W. Goossen,
M. Koch, S. W. Koch, K. Leo, D. A. B. Miller, S. Schmitt-Rink, J. Shah,
and P. Thomas for the very fruitful cooperation.

References

1. S.R. White and L.J. Sham, *Phys. Rev. Lett.* **27**, 879 (1981).
2. G. Bastard, *Phys. Rev. B* **24**, 5693 (1981).
3. G. Bastard, *Phys. Rev. B* **25**, 7584 (1982).
4. M. Altarelli, *Physica.* **117-118B**, 747 (1983).
5. E.E. Mendez and F. Agulló-Rueda, *J. Lumin.* **44**, 223 (1989).
6. E.E. Mendez and G. Bastard, *Physics Today* **June**, 34 (1993).
7. H.M. James, *Phys. Rev.* **76**, 1611 (1949).
8. E.O. Kane, *J. Phys. Chem. Solids* **12**, 181 (1959).
9. G.H. Wannier, *Phys. Rev.* **117**, 432 (1960).
10. J. Callaway, *Quantum theory of the solid state*, 2nd ed. (Academic Press, San Diego, 1991).
11. R.F. Kazarinov and R.A. Suris, *Sov. Phys. Semicond.* **6**, 120 (1972).
12. P.W.A. McIlroy, *J. Appl. Phys.* **59**, 3532 (1986).
13. J. Bleuse, G. Bastard, and P. Voisin, *Phys. Rev. Lett.* **60**, 220 (1988).
14. E.E. Mendez, F. Agulló-Rueda, and J.M. Hong, *Phys. Rev. Lett.* **60**, 2426 (1988).
15. P. Voisin, J. Bleuse, C. Bouche, S. Gaillard, C. Alibert, and A. Regreny, *Phys. Rev. Lett.* **61**, 1639 (1988).
16. F. Agulló-Rueda, E.E. Mendez, and J.M. Hong, *Phys. Rev. B* **40**, 1357 (1989).
17. R.-H. Yan, R.J. Simes, H. Ribot, L.A. Coldren, and A.C. Gossard, *Appl. Phys. Lett.* **54**, 1549 (1989).
18. F. Agulló-Rueda, E.E. Mendez, J.A. Brum, and J.M. Hong, *Surf. Sci.* **228**, 80 (1990).
19. F. Agulló-Rueda, J.A. Brum, E.E. Mendez, and J.M. Hong, *Phys. Rev. B* **41**, 1676 (1990).
20. J. Barrau, K. Khirouni, D.X. Than, T. Amand, M. Brousseau, F. Laruelle, and B. Etienne, *Solid State Commun.* **74**, 147 (1990).
21. P. Tronc, C. Cabanel, J.F. Palmier, and B. Etienne, *Solid State Commun.* **75**, 825 (1990).
22. J. Bleuse, P. Voisin, M. Allovon, and M. Quillec, *Appl. Phys. Lett.* **53**, 2632 (1988).
23. B. Soucail, N. Dupuis, R. Ferreira, P. Voisin, A.P. Roth, D. Morris, K. Gibb, and C. Lacelle, *Phys. Rev. B* **41**, 8568 (1990).
24. K. Fujiwara, H. Schneider, R. Cingolani, and K. Ploog, *Solid State Commun.* **72**, 935 (1989).
25. H. Schneider, H.T. Grahn, K. v. Klitzing, and K. Ploog, *Phys. Rev. Lett.* **65**, 2720 (1990).
26. F. Agulló-Rueda, E.E. Mendez, and J.M. Hong, *Phys. Rev. B* **38**, 12720 (1988).
27. H. Schneider, J. Wagner, K. Fujiwara, and K. Ploog, *Phys. Rev. B* **42**, 11430 (1990).

28. K.H. Schmidt, N. Linder, G.H. Döhler, H.T. Grahn, K. Ploog, and H. Schneider, *Phys. Rev. Lett.* **72**, 2769 (1994).
29. E.E. Mendez, F. Agulló-Rueda, and J.M. Hong, *Appl. Phys. Lett.* **56**, 2545 (1990).
30. F. Cerdeira, C. Vázquez-López, E. Ribeiro, P.A.M. Rodrigues, V. Lemos, M.A. Sacilotti, and A.P. Roth, *Phys. Rev. B* **42**, 9480 (1990).
31. K.W. Goossen, J.E. Cunningham, and W.Y. Jan, *Appl. Phys. Lett.* **59**, 3622 (1991).
32. H. Schneider, A. Fischer, and K. Ploog, *Phys. Rev. B* **45**, 6329 (1992).
33. E. Ribeiro, F. Cerdeira, and A.P. Roth, *Phys. Rev. B* **46**, 12542 (1992).
34. I. Bar-Joseph, K.W. Goossen, J.M. Kuo, R.F. Kopf, D.A.B. Miller, and D.S. Chemla, *Appl. Phys. Lett.* **55**, 340 (1989).
35. H. Schneider, K. Fujiwara, H.T. Grahn, K. v. Klitzing, and K. Ploog, *Appl. Phys. Lett.* **56**, 605 (1990).
36. K.-K. Law, R.H. Yan, J.L. Merz, and L.A. Coldren, *Appl. Phys. Lett.* **56**, 1886 (1990).
37. E. Bigan, M. Allovon, M. Carre, and P. Voisin, *Appl. Phys. Lett.* **57**, 327 (1990).
38. E. Bigan, J.C. Harmand, M. Allovon, M. Carr , A. Carenco, and P. Voisin, *Appl. Phys. Lett.* **60**, 1936 (1992).
39. K.-K. Law, R.H. Yan, L.A. Coldren, and J.L. Merz, *Appl. Phys. Lett.* **57**, 1345 (1990).
40. K. Kawashima, K. Fujiwara, T. Yamamoto, and K. Kobayashi, *Appl. Phys. Lett.* **60**, 1679 (1992).
41. K. Tominaga, M. Hosoda, K. Kawashima, T. Watanabe, and K. Fujiwara, *Appl. Phys. Lett.* **65**, 141 (1994).
42. L.L. Chang and L. Esaki, *Physics Today* Oct., 36 (1992).
43. P. Voisin, G. Bastard, and M. Voos, *Phys. Rev. B* **29**, 935 (1984).
44. H. Chu and Y.-C. Chang, *Phys. Rev. B* **39**, 10861 (1989).
45. C. Weisbuch and B. Vinter, *Quantum semiconductor structures* (Academic Press, New York, 1991).
46. M. Ritze, N.J.M. Horing, and R. Enderlein, *Phys. Rev. B* **47**, 10437 (1993).
47. J. Leo and B. Movaghar, *Phys. Rev. B* **38**, 8061 (1988).
48. J.A. Brum and F. Agulló-Rueda, *Surf. Sci.* **229**, 472 (1990).
49. R.P. Leavitt and J.W. Little, *Phys. Rev. B* **42**, 11784 (1990).
50. G. Bastard, *Wavemechanics applied to semiconductor heterostructures* (Les Editions de Physique, Les Ulis, France, 1988).
51. B. Movaghar, *Semicond. Sci. Technol.* **2**, 185 (1987).
52. J. Leo and A. MacKinnon, *J. Phys. C: Condens. Matter* **1**, 1449 (1989).
53. R.-H. Yan, R.J. Simes, and L.A. Coldren, *IEEE J. Quantum Electron.* **25**, 2272 (1989).
54. S.M. Sze, *Semiconductor devices* (John Wiley, New York, 1985).
55. R.H. Bube, *Photoelectronic properties of semiconductors* (Cambridge University Press, Boston, 1992).

56. F. Agulló-Rueda, H.T. Grahn, A. Fischer, and K. Ploog, *Phys. Rev. B* **45**, 8818 (1992).

57. E.E. Mendez, G. Bastard, L. L. Chang, L. Esaki, H. Morkoç, and R. Fischer, *Phys. Rev. B* **26**, 7101 (1982).

58. G. Bastard, E.E. Mendez, L.L. Chang, and L. Esaki, *Phys. Rev. B* **28**, 3241 (1983).

59. A.M. Fox, D.A.B. Miller, J.E. Cunningham, W.Y. Jan, C.Y.P. Chao, and S.L. Chuang, *Phys. Rev. B* **46**, 15365 (1992).

60. M. Dignam and J.E. Sipe, *Phys. Rev. Lett.* **64**, 1797 (1990).

61. D.M. Whittaker, *Superlattices Microstruct.* **7**, 375 (1990).

62. M. Dignam and J.E. Sipe, *Phys. Rev. B* **43**, 4097 (1991).

63. K. Kawashima, T. Yamamoto, K. Kobayashi, and K. Fujiwara, *Phys. Rev. B* **47**, 9921 (1993).

64. A. Chomette, B. Lambert, B. Deveaud, F. Clerot, A. Regreny, and G. Bastard, *Europhys. Lett.* **4**, 461 (1987).

65. S. Adachi, *J. Appl. Phys.* **58**, R1 (1985).

66. D.M. Whittaker, *Phys. Rev. B* **41**, 3238 (1990).

67. F.L. Lederman and J.D. Dow, *Phys. Rev. B* **13**, 1633 (1976).

68. D.E. Aspnes, in *Handbook on Semiconductors*, edited by M. Balkanski and T. S. Moss (North-Holland, Amsterdam, 1980), Vol. 2, pp. 109.

69. Y. Hamakawa, F.A. Germano, and P. Handler, *Phys. Rev.* **167**, 703 (1968).

70. D.M. Whittaker, M.S. Skolnik, G.W. Smith, and C.R. Whitehouse, *Phys. Rev. B* **42**, 3591 (1990).

71. M.K. Saker, D.M. Whittaker, M.S. Skolnik, M.T. Emeny, and C.R. Whitehouse, *Phys. Rev. B* **43**, 4945 (1991).

72. A.J. Shields, P.C. Klipstein, M.S. Skolnick, G.W. Smith, and C.R. Whitehouse, *Phys. Rev. B* **42**, 5879 (1990).

73. M. Nakayama, I. Tanaka, H. Nishimura, K. Kawashima, and K. Fujiwara, *Phys. Rev. B* **44**, 5935 (1991).

74. H. Schneider, K. Kawashima, and K. Fujiwara, *Phys. Rev. B* **44**, 5943 (1991).

75. I. Tanaka, M. Nakayama, H. Nishimura, K. Kawashima, and K. Fujiwara, *Phys. Rev. B* **46**, 7656 (1992).

76. R.P. Leavitt and J.W. Little, *Phys. Rev. B* **41**, 5174 (1990).

77. F. Agulló-Rueda, E.E. Mendez, H. Ohno, and J.M. Hong, *Phys. Rev. B* **42**, 1470 (1990).

78. Y. Tokuda, K. Kanamoto, Y. Abe, and N. Tsukada, *Phys. Rev. B* **43**, 7170 (1991).

79. J.L. Bradshaw, R.P. Leavitt, J.T. Pham, and F.J. Towner, *Phys. Rev. B* **49**, 1882 (1994).

80. H. Ohno and E.E. Mendez, J.A. Brum, J.M. Hong, F. Agulló-Rueda, L.L. Chang, *Phys. Rev. Lett.* **64**, 2555 (1990).

81. F. Agulló-Rueda, A. D'Intino, K.H. Schmidt, G.H. Döhler, H.T. Grahn, and K. Ploog, *Europhys. Lett.* **23**, 283 (1993).

82. B. Movaghar, *Semicond. Sci. Technol.* **3**, 908 (1988).
83. F. Claro, M. Pacheco, and Z. Barticevic, *Phys. Rev. Lett.* **64**, 3058 (1990).
84. R. Ferreira, B. Soucail, P. Voisin, and G. Bastard, *Phys. Rev. B* **42**, 11404 (1990).
85. A. Alexandrou, E.E. Mendez, and J.M. Hong, *Phys. Rev. B* **44**, 1934 (1991).
86. M. Dignam and J.E. Sipe, *Phys. Rev. B* **45**, 6819 (1992).
87. A. Alexandrou, M.M. Dignam, E.E. Mendez, J.E. Sipe, and J.M. Hong, *Phys. Rev. B* **44**, 13124 (1991).
88. F. Agulló-López, J.M. Cabrera, and F. Agulló-Rueda, *Electrooptics* (Academic Press, New York, 1994).
89. K. Kawashima, K. Fujiwara, T. Yamamoto, M. Sigeta, and K. Kobayashi, *Surf. Sci.* **267**, 643 (1992).
90. F. Bloch, *Z. Phys.* **52**, 555 (1928).
91. C. Kittel, *Quantum Theory of Solids* (Wiley, New York, 1963).
92. J. Zak, *Solid State Physics* (Academic Press, New York, 1972).
93. A. Nenciu and G. Nenciu, *Phys. Lett.* **78**, 101 (1980).
94. J.B. Krieger and G.J. Iafrate, *Phys. Rev. B* **33**, 5494 (1986).
95. V.W. Houston, *Phys. Rev.* **57**, 184 (1940).
96. C. Zener, *Proc. R. Soc.* A145, 523 (1934).
97. P. Roblin and M.W. Muller, *Semicond. Sci. Technol.* **1**, 218 (1986).
98. G. Nenciu, *Rev. Mod. Phys.* **63**, 91 (1991).
99. W. Franz, *Z. Naturforschg.* 13a, 484 (1958).
100. L.V. Keldysh, *Sov. Phys. JETP* **34**, 788 (1958).
101. F. Beltram, F. Capasso, D.L. Sivco, A.L. Hutchinson, S. Chu, and A.Y. Cho, *Phys. Rev. Lett.* **64**, 3167 (1990).
102. D.E. Aspnes and A.A. Studna, *Phys. Rev. B* **7**, 4605 (1973).
103. C. Coriasso, D. Campi, C. Cacciatore, C. Alibert, S. Gaillard, B. Lambert, and A. Regreny, *Europhys. Lett.* **16**, 591 (1991).
104. G. Fasol, W. Hackenberg, H.P. Hughes, K. Ploog, E. Bauser, and H. Kano, *Phys. Rev. B* **41**, 1461 (1990).
105. E.O. Göbel, *Festkörperprobleme/Advances in Solid State Physics* **30**, 269 (1990).
106. see, e.g., M. Koch, D. Weber, J. Feldmann, E.O. Göbel, T. Meier, A. Schulze, P. Thomas, S. Schmitt-Rink, and K. Ploog, *Phys. Rev. B* **47**, 1532 (1993).
107. S.M. Zakharov and E.A. Manykin, *Izv. Akad. Nauk SSSR* **37**, 2171 (1973).
108. G. von Plessen and P. Thomas, *Phys. Rev. B* **45**, 9185 (1992).
109. L. Allen and J.H. Eberly, *Optical Resonances and Two-Level Systems* (Dover Publications, New York, 1975).
110. J. Feldmann, K.Leo, J. Shah, D.A.B. Miller, J.E. Cunningham, T. Meier, G. von Plessen, A. Schulze, P. Thomas, and S. Schmitt-Rink, *Phys. Rev. B* **46**, 7252 (1992).
111. K. Leo, P.H. Bolivar, F. Brüggemann, R. Schwedler, and K. Köhler, *Solid State Commun.* **84**, 943 (1992).
112. A.M. Bouchard and M. Luban, *Phys. Rev. B* **47**, 6815 (1993).

113. L. Esaki and R. Tsu, *IBM J. Dev.* **14**, 61 (1970).

114. C. Waschke, H.G. Roskos, R. Schwedler, K. Leo, H. Kurz, and K. Köhler, *Phys. Rev. Lett.* **70**, 3319 (1993); H.G. Roskos, C. Waschke, R. Schwedler, P. Leisching, Y. Dhaibi, H. Kurz, and K. Köhler, *Superlattices Microstruct.*, in press (1994).

115. J. Kuhl, A. Honold, L. Schultheis, and C.W. Tu, *Festkörperprobleme/ Advances in Solid State Physics* **29**, 157 (1989).

116. G. von Plessen, T. Meier, J. Feldmann, E.O. Göbel, P. Thomas, K.W. Goossen, J.M. Kuo, and R.F. Kopf, *Phys. Rev. B* **49**, 14058 (1994).

117. J. Feldmann, *Festkörperprobleme/Advances in Solid State Physics* **32**, 81 (1992).

118. I. Tanaka, M. Nakayama, H. Nishimura, K. Kawashima, and K. Fujiwara, *Phys. Rev. B* **48**, 2787 (1993).

119. R.P. Leavitt, J.L. Bradshaw, and F.J. Tower, *Phys. Rev. B* **44**, 11266 (1991).

120. S. Schmitt-Rink, D.S. Chemla, W.H. Knox, and D.A.B. Miller, *Opt. Lett.* **15**, 60 (1990).

121. H. Haug and S.W. Koch, *Quantum Theory of the Optical and Electronic Properties of Semiconductors* (World Scientific, Singapore, 1990).

122. J. Feldmann, T. Meier, G. von Plessen, M. Koch, E.O. Göbel, P. Thomas, G. Bacher, C. Hartmann, H. Schweizer, W. Schäfer, and H. Nickel, *Phys. Rev. Lett.* **70**, 3027 (1993).

123. G. von Plessen, T. Meier, M. Koch, J. Feldmann, E.O. Göbel, P. Thomas, S.W. Koch, K.W. Goossen, J.M. Kuo, and R.F. Kopf, to be published.

124. G. von Plessen, PhD-thesis, Philipps-University of Marburg, Marburg 1994.

125. K. Leo, M. Wegener, J. Shah, D.S. Chemla, E.O. Göbel, T.C. Damen, S. Schmitt-Rink, and W. Schäfer, *Phys. Rev. Lett.* **65**, 1340 (1990).

126. M. Wegener, D.S. Chemla, S. Schmitt-Rink, and W. Schäfer, *Phys. Rev. A* **42**, 5675 (1990).

127. M. Lindberg, R. Binder, and S.W. Koch, *Phys. Rev. A* **45**, 1865 (1992).

128. D.S. Kim, J. Shah, T.C. Damen, W. Schäfer, F. Jahnke, S. Schmitt-Rink, and K. Köhler, *Phys. Rev. Lett.* **69**, 2725 (1992).

129. S. Weiss, M.A. Mycek, J.Y. Bigot, S. Schmitt-Rink, and D.S. Chemla, *Phys. Rev. Lett.* **69**, 2685 (1992).

130. T. Meier, G. von Plessen, P. Thomas, and S.W. Koch, *Phys. Rev. Lett.* **73**, 902 (1994).

CHAPTER 4

RESONANT TUNNELING

by HOLGER GRAHN

4.1. Introduction

Tunneling is a quantum mechanical phenomenon, which does not have a classical analogy. The first application of resonant tunneling was the description of splitting of the energy levels in a molecule consisting of two atoms by Hund.[1] Further investigations dealt with thermal emission of electrons from metals[2] and the description of the α-decay of radioactive nuclei.[3,4] The first application of the concept of tunneling to semiconductors and insulators was performed by Zener,[5] who described the electrical breakdown of dielectrics by interband tunneling. In the 1950s the model was used to describe the electrical transport characteristic of the Esaki tunneling diode.[6] A recent review of resonant tunneling in semiconductors is given in Ref. 7.

Artificial superlattices represent a model system for the investigation of quantum phenomena as was already pointed out by Esaki and Tsu in their pioneering work.[8] In particular the realization of superlattices using semiconductor materials with very different band gaps, e.g., GaAs and AlAs, is ideal to study resonant tunneling between quantum wells. Due to the spatial confinement of the electrons or holes within a quantum well, resonant tunneling between two-dimensional systems can be investigated. The superlattice or coupled multi quantum well system has two advantages in comparison with double barrier and aysmmetric double quantum well structures. First, tunneling takes place between identical wells. Second, the effects of the contacts can be completely neglected, since the current is dominated by the transport within the superlattice. A magnetic field parallel to the superlattice direction, i.e. perpendicular to the layers, reduces the dimensionality to zero, since the in-plane free electron states are quantized into Landau levels. Consequently, resonant tunneling between zero-dimensional states can be investigated in this configuration.

In the previous two chapters the effect of an electric field parallel to the superlattice direction was discussed in the low-field regime, i.e., miniband transport, and intermediate-field regime, i.e., Wannier-Stark localization and Bloch oscillations. Only the lowest miniband was taken into account. In this chapter we will consider the superlattice in the high-field regime, when the states are completely localized again within a single well, i.e., the quasi three-dimensional miniband is transformed into two-dimensional subbands. However, due to the thin barriers, the superlattice represents in the high-field regime a *coupled* multi quantum well (MQW) system rather than an uncoupled MQW system. The coupling of the lowest subband with a higher subband in the adjacent well leads to resonant tunneling in this field regime. In this chapter we will be concerned with the consequences of resonant tunneling on the properties of the superlattice in the low carrier density limit. For large carrier densities the field distribution within the superlattice is strongly affected by resonant tunneling resulting in the formation of electric-field domains. This subject will be discussed extensively in Chapter 5.

As early as 1971 possible application of resonant tunneling in semiconductor superlattices was discussed with regard to the amplification of electromagnetic waves.[9] Due to the injection of carriers by resonant tunneling into a higher subband, a non-thermal carrier distribution can be achieved.[10] If the tunneling process is faster than the intersubband relaxation, inversion between the first two subbands is possible. While intersubband inversion within one well has not been reported so far, it was achieved in a triple well structure with very narrow barriers leading to the invention of the quantum cascade laser.[11]

4.2. Theoretical background

Resonant tunneling in semiconductor heterostructures has been a very active area of research within the last ten years. Since we are only interested in the coupling of adjacent wells, we can consider a symmetric double quantum well as the basic unit of the superlattice. This system is analogous to the NH_3 molecule, in which two bound states are separated by a potential barrier. The probability of finding the system in either potential well becomes the same at resonance. Therefore, in a time-dependent description the system is oscillating back and forth between the two potential wells. In the case of the NH_3 molecule this is demonstrated by the emission of electromagnetic radiation with a characteristic frequency ω, which is determined by the level splitting ΔE^{res} at resonance, i.e.,[12]

$$\omega = \frac{\Delta E^{res}}{\hbar}. \qquad (4.1)$$

For a symmetric double quantum well in an applied electric field the level splitting can only be determined by a numerical solution of Schrödinger's equation. However, the formula in Eq. 4.1 holds not only for the coupling of the ground states in the symmetric double quantum well system, but also for the coupling of the ground state in one well with an excited state in the other well. For a level splitting of 5 meV, an oscillation period of 0.83 ps is obtained from Eq. 4.1. The coherent tunneling time is assumed to be one half of the oscillation period.[13] If the level splitting is smaller than the width of the respective energy levels, the coherence of the tunneling process is destroyed. In this case, the complete resonant tunneling process is referred to as incoherent or sequential, i.e., an elastic or inelastic scattering process occurs during the tunneling process. We will be mainly concerned with the limit of incoherent tunneling, since the observation of coherent tunneling requires a subpicosecond time resolution as discussed at the end of the previous chapter.

The electric field strength for resonant tunneling between different subbands has also to be determined by a numerical solution of Schrödinger's equation, since the eigenenergies of each well depend on the electric field via the quantum-confined Stark effect (QCSE).[14] However, for smaller well widths the QCSE can be neglected to first approximation. In this case the resonance field strengths are directly determined by the subband spacing within a single well

$$F_{ij}^{res} = \frac{E_j - E_i}{e\,d}, \qquad (4.2)$$

where E_i and E_j denote the energy of the two involved subbands, d the superlattice period, and e the elementary charge. In most cases $i = 1$. Since each subband experiences a different Stark shift, the resonance field strengths have to be calculated self-consistently. Typical values for F^{res} are

10 to 400 kVcm^{-1}. In order to resolve a tunneling resonance, the subband spacing has to be above the level broadening due to impurities and interface roughness, which is typically of the order of 1 meV.

After tunneling into a higher subband the carriers can either scatter down into a lower subband via intersubband scattering (IS) or tunnel non-resonantly (NT) into the next well. In a superlattice with a rectangular potential profile it is impossible to achieve for a given field strength a resonance between three wells, since the subband spacing increases with higher subband index. The typical transport process is shown schematically in Fig. 4.1 for tunneling from the first into the second subband in the adjacent well.

The intersubband scattering time strongly depends on the subband spacing. If the subband spacing is smaller than the optical phonon energy

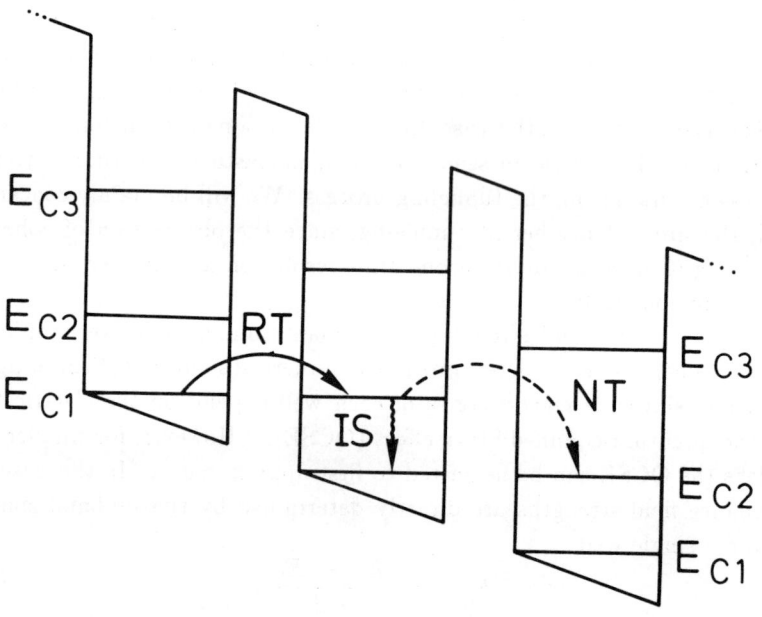

Fig. 4.1. Schematic diagram of carrier injection by resonant tunneling (RT) between conduction band levels C1 and C2 as well as subsequent processes such as intersubband scattering (IS) and non-resonant tunneling (NT) including an inelastic scattering process.

$\hbar\omega_{LO}$, e.g., in GaAs $\hbar\omega_{LO} = 36$ meV, the intersubband scattering time τ_{IS} becomes rather large by reaching several hundreds of picoseconds.[15-17] However, if the subband spacing is larger than $\hbar\omega_{LO}$, τ_{IS} is strongly reduced to values below 1 ps.[10,16,18-23]

The occupation of the second subband by resonant tunneling can be estimated from a simple rate equation model assuming that the tunneling process is incoherent. The carrier density in the first subband is reduced by tunneling and increased by intersubband relaxation, i.e.,

$$\frac{dn_1^i}{dt} = -\frac{n_1^i}{\tau_{RT}} + \frac{n_2^i}{\tau_{IS}} + F_1(t) , \qquad (4.3)$$

where n_1^i (n_2^i) denotes the occupation of the first (second) subband in the ith well, τ_{RT} the incoherent tunneling time, and $F_1(t)$ the optical generation rate of the carriers for pulsed excitation. For the second subband tunneling increases the occupation and intersubband scattering reduces it, i.e.,

$$\frac{dn_2^i}{dt} = -\frac{n_2^i}{\tau_{IS}} + \frac{n_2^{i-1}}{\tau_{RT}} + F_2(t) . \qquad (4.4)$$

$F_2(t)$ denotes the generation rate in the second subband. We have neglected the non-resonant channel for tunneling out of the second subband. Assuming that the carrier density in adjacent wells is almost the same, i.e., $n_1^{i-1} \approx n_1^i$, the steady-state solution of these two equations becomes very simple

$$\frac{n_2^i}{n_1^i} = \frac{\tau_{IS}}{\tau_{RT}} . \qquad (4.5)$$

Assuming the same initial carrier distribution for the first and second subband, the ratio of the total occupation of the second with respect to the first subband is equal to the one given in Eq. 4.5. In order to achieve a large population by resonant tunneling, it is necessary to increase the intersubband scattering time with regard to the tunneling time. Either the tunneling time has to become very short, i.e., less than a few hundred femtoseconds, or the intersubband scattering time has to be increased by reducing the subband spacing below the optical phonon energy.

4.3. Experimental evidence for resonant electron tunneling

The first observation of resonant tunneling was reported by Esaki and Chang.[24] The conductance of a doped, weakly coupled GaAs/AlAs superlattice exhibited periodic oscillations with a spacing related to the spacing of the first and second electronic subband. However, this oscillations are due to domain formation and will be discussed in Chapter 5.

In order to identify tunneling resonances, the experiments are usually performed on undoped superlattices embedded in a p-i-n diode in order to achieve a homogeneous field over the superlattice region. In reverse bias no carriers are injected into the intrinsic layer, and the carrier density can solely be controlled by photoexcitation. The superlattice transport properties can be studied either by photocurrent-voltage characteristics or time-of-flight experiments. A typical band edge distribution of a superlattice within a p-i-n diode without any externally applied voltage is shown in Fig. 4.2 The applied electric field F_{ap} is related to the applied voltage V_{ap} through

$$F_{ap} = (V_{bi} - V_{ap})/L_{in} , \qquad (4.6)$$

where V_{bi} denotes the built-in voltage (typically at low temperature about

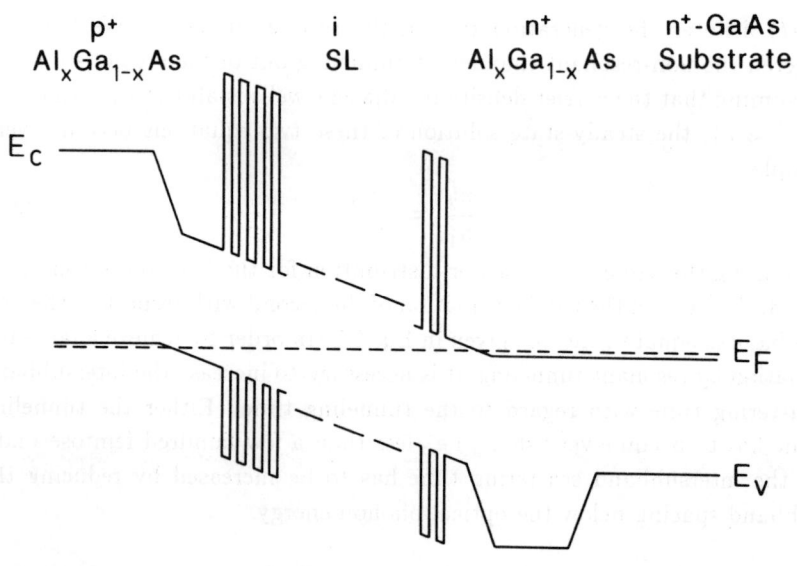

Fig. 4.2. Schematic diagram of the band edge distribution within a p-i-n diode without any externally applied voltage. The superlattice consists of GaAs wells and AlAs barriers. The band gaps of GaAs and AlAs are drawn to scale. E_F denotes the Fermi level, E_C and E_V the conduction and valence band edge, respectively.

1.5 V) and L_{in} the width of the intrinsic region of the p-i-n diode (typically of the order of 1 μm).

4.3.1. *Photocurrent-voltage characteristics*

The first observation of resonant tunneling between the lowest and a higher subband in a semiconductor superlattice was reported by Capasso et al.[25] The photocurrent-voltage characteristic of an undoped $In_{0.53}Ga_{0.47}As$-$In_{0.52}Al_{0.48}As$ superlattice with 35 periods and well and barrier widths of 13.9 nm was recorded at different temperatures showing clearly the tunneling resonances $C1 \rightarrow C2$ and $C1 \rightarrow C3$ between adjacent wells. The experimental result is shown in Fig. 4.3. When the voltage difference between the two peaks is divided by the number of periods, a value of 140 mV

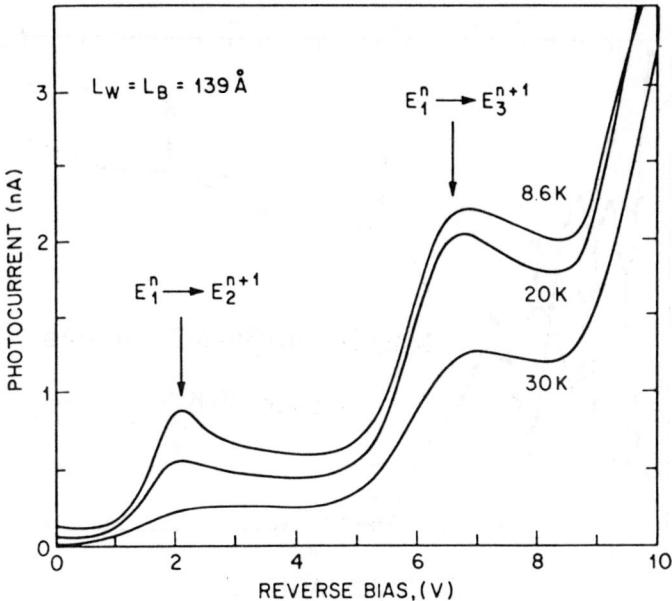

Fig. 4.3. Photocurrent-voltage characteristic in a superlattice with 35 periods, 13.9 nm $In_{0.53}Ga_{0.47}As$ wells, and 13.9 nm $In_{0.52}Al_{0.48}As$ barriers at different temperatures. The excitation energy was 1.959 eV. The arrows indicate the tunneling resonances $C1 \rightarrow C2$ and $C1 \rightarrow C3$ (from Ref. 25).

is obtained in excellent agreement with the calculated subband spacing of the second and third electronic subband of 143 meV. This observation was the first direct evidence of sequential resonant tunneling as described in the previous section. The authors further analyzed the width of the resonances using the model proposed by Kazarinov and Suris[9] estimating a transverse momentum relaxation time of 0.1 ps, which describes the relaxation of the phase difference between the states involved in the resonant tunneling process.

The first observation of sequential resonant tunneling in superlattices with GaAs wells separated by AlGaAs barriers was reported by Furuta et al.[26] The sequential resonant tunneling characteristic of GaAs/AlAs super-lattices was investigated by Tarucha et al.[27,28] using photocurrent-voltage characteristic and photoluminescence (PL) spectroscopy. In Fig. 4.4 a com-

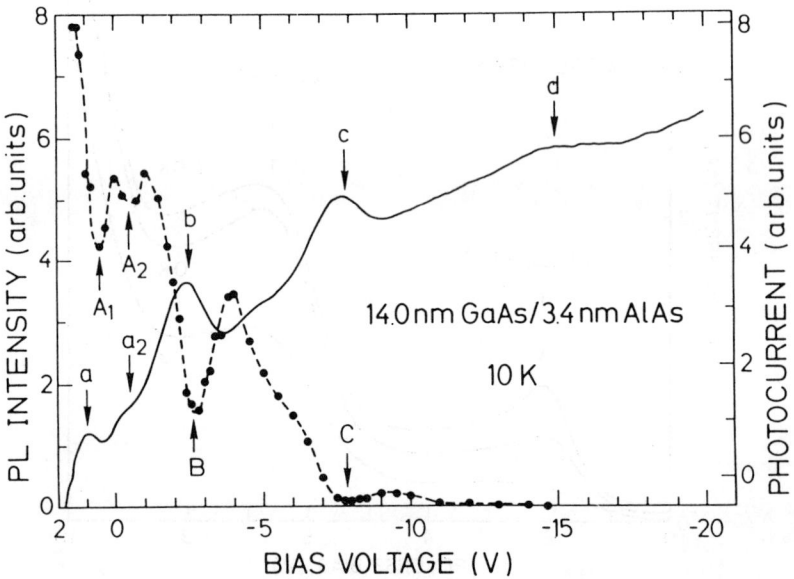

Fig. 4.4. Static photocurrent (solid line) and spectrally integrated photoluminescence intensity (solid circles) vs bias voltage in a superlattice with 14 nm GaAs wells, 3.4 nm AlAs barriers, and 50 periods measured under the same cw excitation conditions at 10 K. The dashed line connecting the solid circles is a guide to the eye (from Ref. 28).

parison of the static photocurrent and photoluminescence intensity vs applied voltage is shown for GaAs/AlAs superlattice with 14 nm GaAs wells, 3.4 nm AlAs barriers, and 50 periods. Four strong maxima in the photocurrent correlate with minima in the photoluminescence intensity. Resonant tunneling from the first into the second, third, and fourth electronic subband in the next well is responsible for the maxima labeled *b*, *c*, and *d*, respectively. The applied voltages for the peaks *b*, *c*, and *d* are in good agreement with the expected resonance fields determined from the calculated subband spacing. The PL intensity exhibits minima at the voltages corresponding to peak *b* and *c*. This can be understood in terms of an enhanced transport probability at the resonance voltages leading to a reduction of radiative recombination. The peaks *a* and a_2 were not assigned to any resonances. However, peak *a* could be due to miniband transport and peak a_2 due to phonon-assisted tunneling. We will return to the subject of phonon-assisted tunneling in the next section.

4.3.2. *Time-of-flight experiments*

The static photocurrent-voltage characteristic is obtained under continuous illumination. A quasi-stationary state between generation, transport and recombination of the photo-excited carriers is achieved. A partial screening of the applied electric field can occur, and the tunneling resonances become much weaker or disappear in the photocurrent-voltage characteristic. It is therefore better to use a time-resolved technique such as the time-of-flight method for the investigation of tunneling resonances. A short laser pulse excites electron-hole pairs near the top p-contact in the superlattice. Under the applied electric field electrons drift through the superlattice to the n-contact at the back, while the holes move on a much longer time scale to the front p-contact due to their heavier mass. While the electrons drift through the superlattice, a displacement current is observed, which vanishes after all electrons have reached the back contact. However, the transient depends strongly on the applied voltage, since at resonance the electrons can move faster through the superlattice than for non-resonant field strengths. Since the time-integrated current,. i.e., the photogenerated charge, is a constant, the transients become shorter in time and their amplitude larger at resonance. It is therefore sufficient to either plot the peak photocurrent vs the applied voltage or the measured transport time. Since the peak photocurrent can be directly measured, while the transport time has to be obtained by fitting the transients to a model, we will use the peak photocurrent as a measure for the superlattice transport probability.

This time-of-flight method has been successfully applied to study carrier transport through superlattices.[27–32] Resonant tunneling in coupled multi quantum well system has also been investigated by measuring the field dependence of the rise time of the differential absorption signal.[33] In Fig. 4.5 a typical result is shown for a superlattice with 18.4 nm GaAs wells, 1.4 nm AlAs barriers, and 40 periods. In addition to the strong tunneling resonances between the first and the second, third and fourth electronic subband at −0.9, −4.5, and −9.8 V, respectively, additional, but weaker resonances are observed between the $C1 \rightarrow C3$ and $C1 \rightarrow C4$ resonances at −6.6 and −7.7 V and also above the $C1 \rightarrow C4$ resonance at −12.7 V. The first two resonances are due to resonant tunneling from the first into the third subband in the adjacent well in conjunction with the emission

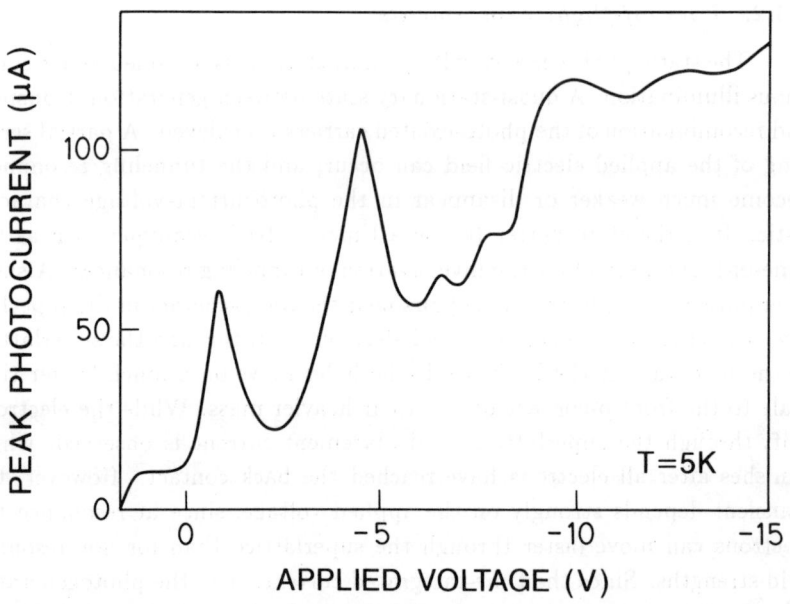

Fig. 4.5. Peak photocurrent vs applied voltage in a superlattice with 18.4 nm GaAs wells, 1.4 nm AlAs barriers, and 40 periods at 5 K. The excitation energy was 1.851 eV.

of a longitudinal optical (LO) phonon of GaAs ($\hbar\omega_{LO}$=36 meV) and AlAs ($\hbar\omega_{LO}$=50 meV), respectively. The third additional resonance is assigned to resonant tunneling between the first and fourth electronic subband in conjunction with the emission of a GaAs LO phonon. The phonon assisted tunneling peaks between the first two resonances are probably masked by the relatively broad $C1 \rightarrow C3$ resonance.

The transport time τ_{res} through the whole superlattice can be estimated from the photogenerated charge Q_0 divided by the peak photocurrent. For the sample in Fig. 4.5 we obtain 0.74 and 0.39 ns for the first two large tunneling resonances. Dividing this time by the number of periods, an estimate of the average transport time between adjacent wells is obtained. For the $C1 \rightarrow C2$ resonance the result is 14.8 ps, while for the $C1 \rightarrow C3$ resonance the resonant transfer time is 7.8 ps. Both times are about one order of magnitude larger than the calculated coherent resonant tunneling times. The resonance condition cannot be fulfilled in real structures for all carriers at the same time due to local fluctuations of the well thickness and field inhomogeneities. The smaller the width of the resonance width, i.e., ΔE^{res}, the larger these effects and the tunneling time increases. We conclude that in these structures the tunneling process is incoherent and dominated by sequential tunneling. Coherent resonant tunneling was recently observed in time-resolved four-wave mixing and terahertz emission experiments on asymmetric double quantum well structures.[34,35]

So far we only considered the resonant tunneling process at low temperatures, when only the lowest subband injects carriers into a higher subband in the adjacent well. In Fig. 4.6 the temperature dependence of the peak photocurrent-voltage characteristic is shown between 8.5 and 295 K. For temperatures of 150 K and above an additional peak appears between the $C1 \rightarrow C2$ and $C1 \rightarrow C3$ resonances. Furthermore, the LO phonon assisted tunneling peaks dissappear and between the $C1 \rightarrow C3$ and $C1 \rightarrow C4$ resonance a new peak emerges. A careful analysis of the peak positions taking into the shift in the built-in voltage with temperature identifies these resonances to be due to elastic tunneling out of the thermally populated second subband $C2$ into $C3$ and $C4$, respectively. Assuming thermal equilibrium, the occupation of the second subband n_2 can be estimated from

$$n_2 = n_1 \, exp\left(-\frac{E_2 - E_1}{k_B \, T}\right) , \qquad (4.7)$$

where T denotes the temperature and k_B Boltzmann's constant. At 295 K the occupation ratio $n_2/n_1 \approx 0.2$, which corresponds to a relative occupation ratio of the second subband $n_2/(n_1 + n_2)$ of 17%. This idenfication has been confirmed by measuring the luminescence spectrum up to 295 K. From 150 K on the transition between the second electronic subband $C2$ and the second heavy-hole subband $H2$ becomes visible.[36]

In addition to the observation of tunneling out of a thermally excited subband, the competition between thermally induced resonant tunneling and phonon-assisted tunneling is clearly seen in Fig. 4.6. At low temperatures phonon-assisted tunneling channels, i.e., inelastic tunneling, can be

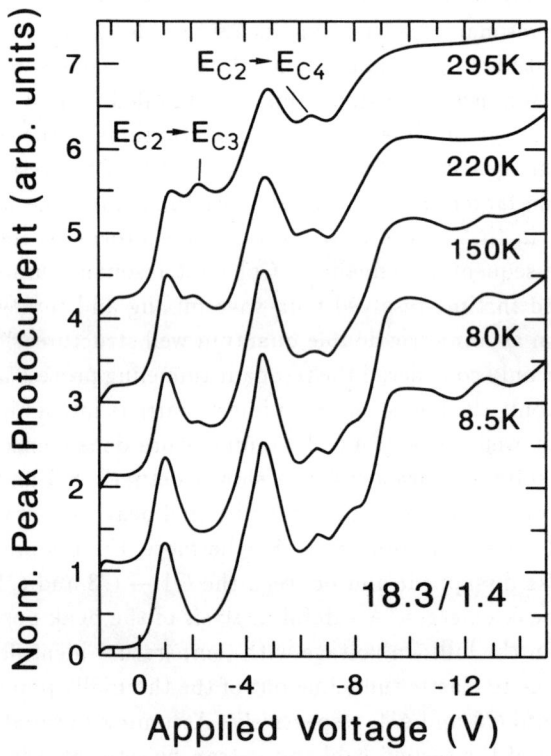

Fig. 4.6. Peak photocurrent vs applied voltage in a superlattice with 18.4 nm GaAs wells, 1.4 nm AlAs barriers, and 40 periods at different temperatures. The excitation energy was 1.851 eV.

present. If the subband spacing is much larger than the thermal energy at room temperature, no significant thermal population of the second subband is achieved and the inelastic tunneling are always seen.[36] However, for wider wells, the subband spacing can become comparable to the thermal energy and the direct tunneling channel out of a thermally excited subband will dominate over the indirect tunneling channel.

4.3.3. *Non-thermal occupation of higher subbands by resonant tunneling*

The non-thermal occupation of higher subbands via resonant tunneling can be directly observed by detecting the intersubband transitions in infrared emission experiments[37,38] or the interband transitions involving higher subbands in photoluminescence[10,39-40] and electroluminescence experiments.[41-43] In order to study the intersubband transitions, it is necessary that the subband spacing is well below the optical phonon energy to achieve a sufficient population of a higher subband.

4.3.3.1. *Infrared emission experiments*

The infrared emission experiments reported by Helm et al.[37,38] clearly show the emission of far-infrared light corresponding to transitions from $C2$ to $C1$ and from $C3$ to $C2$. Since the intersubband emission is polarized parallel to the layers, the experiments have to be performed by detecting the emission from the edge of the sample or by evaporating a grating coupler on top of the sample, which produces a component of the polarization vector perpendicular to the layers. The experiments by Helm et al. used the grating coupler, since they were not able to observe any signal from the edge of the sample. In Fig. 4.7 a typical experimental result of an infrared emission experiment is shown. The experiment was performed at low temperatures (10-20 K) using a broadband Si-bolometer for detection and a magnetic-field tunable InSb filter to analyze the spectrum. Minima in the signal correspond to maxima in the emission spectrum. The observed intersubband emission transitions are identified by comparing the energies with the calculated subband spacing. The strongest transitions are $C2$ to $C1$ and $C3$ to $C2$. However, for even higher voltages also the $C4$ to $C3$ transition is observed, although it is rather weak, since the fourth subband is already separated from the ground state by more than the optical phonon energy.

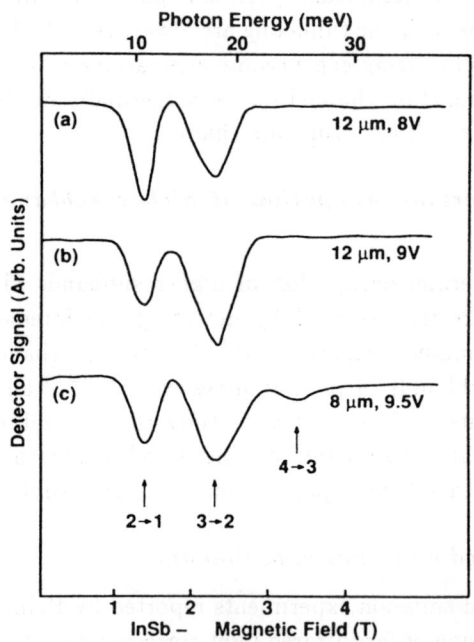

Fig. 4.7. Detector signal vs magnetic field of the InSb filter and corresponding photon energy at 10-20 K in a superlattice with 35 nm GaAs wells, 10 nm Al$_{0.3}$Ga$_{0.7}$As barriers, and 60 periods for different grating periods at the bias voltages indicated. The observed intersubband transitions are indicated (from Ref. 37).

The experiments were actually performed by electrically injecting carries into the fifth or sixth subband. At the resonance voltages for tunneling into the second or third subband, no infrared emission signal could be detected. The superlattice was not intentionally doped, but sandwiched between n$^+$-contacts. Due to the rather thick barriers, the I-V characteristics clearly showed the signature of electric-field domain formation, which will be discussed in Chapter 5. In order to increase the population at the resonance voltage and reduce the tendency for electric-field domain formation, thinner barriers are necessary.

4.3.3.2. *Photoluminescence spectroscopy*

The carrier density within the superlattice can be very well controlled by using a reverse biased p-i-n diode under photoexcitation. In order to achieve a sufficiently large population for detection at the corresponding resonance voltage, rather thin barriers have to be used. The recombination of electrons in a higher conduction subband with holes in the ground state ($H1$) is in infinitely deep wells a forbidden transition at zero electric field. However, due to the breakdown of the symmetry of the wave function, when an electric field is applied perpendicular to the layer, these transitions gain oscillator strength with increasing field. The intensity of an interband transition I_{ij} is proportional to the product of the occupation numbers n_i and p_j of the respective subbands and to the corresponding oscillator strengths O_{ij}

$$I_{ij} \propto n_i \times p_j \times O_{ij} . \tag{4.8}$$

The oscillator strength can be determined by calculating the overlap integral between the envelope wave functions $\Phi(x)$

$$O_{ij} = | \int\limits_{-\infty}^{+\infty} \Phi_i^C(x) \times \Phi_j^{H,L}(x) dx |^2 . \tag{4.9}$$

Taking the envelope wave functions $\Phi_i^C(x)$ and $\Phi_j^{H,L}(x)$ from the numerical solution of Schrödinger's equation, we determine O_{ij} for different subband indices as a function of the applied electric field. The results for the strongest interband transitions, i.e., the largest overlap integrals, are shown in Fig. 4.8 for a quantum well with a finite barrier height, i.e., a 21 nm GaAs well sandwiched between AlAs barriers. The envelope functions are normalized to unity resulting in a maximum value of 1 for the overlap integral. As mentioned above, only transitions with equal subband index exhibit a sizeable oscillator strength for vanishing electric field, e.g., $C1H1$, $C2H2$, etc. With increasing field O_{C1H1} decreases continuously to about 0.1 for an applied field of about 40 kV/cm. At the same time the transition probability of the $C2H1$ transition increases to a maximum value of 0.6 at about 32 kV/cm and then decreases again. The other transition continuously increase in the displayed field range. At the $C1 \rightarrow C3$ resonance the $C2H1$, $C3H1$, and $C1H3$ overlap integrals are already larger than the one of the $C1H1$ transition. This implies that efficient emission from these transitions can be expected if the involved conduction and valence subbands are significantly occupied. The transitions with equal subband index like $C2H2$ and $C3H3$ are less likely to be observed due to the simultaneous low occupation of higher conduction *and* higher valence subbands in the same well and decreasing overlap integrals of the $CnHn$ transitions for larger electric fields. The transitions shown in Fig. 4.8 are therefore the most probable ones involving a higher subband in one band and the ground state in the other. Mixing between heavy-hole and light-hole states has been neglected. Samples with very different subband spacing have been investigated. In Fig. 4.9 photoluminescence spectra for different applied electric fields are shown for a superlattice with 12.3 nm GaAs wells, 2.2 nm AlAs barriers,

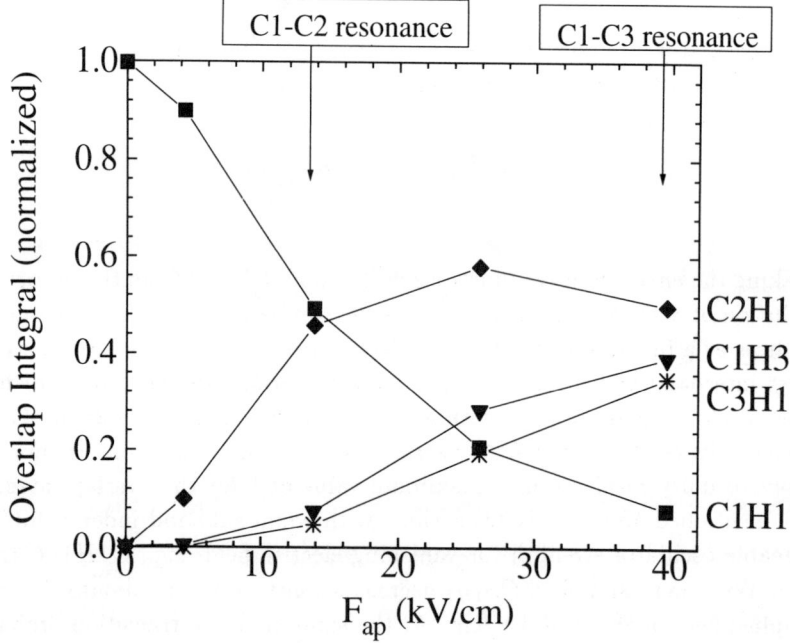

Fig. 4.8. Overlap integrals of the normalized electron and hole wave functions vs electric field for a single GaAs quantum well with a width of 21 nm sandwiched between AlAs barriers. The involved subbands are indicated. The solid lines are only a guide to the eye.

and 50 periods. The first and second electronic subband are separated by 80 meV at zero electric field, which is much more than the optical phonon energy. Nevertheless, the non-thermal occupation of the second and third electronic subband via resonant tunneling can be directly seen in Fig. 4.9. While in Fig.4.9(a) carriers were excited with a photon energy of 1.90 eV far above the second and third electronic subband, the higher subband PL is still observed, when the laser energy is tuned between the first and second subband. Therefore, the observed PL signal is not due to photoexcited hot carriers, but rather due to hot carriers that are injected by resonant tunneling. In order to quantitatively determine the occupation of the second subband, the oscillator strength has to be determined as a function of the applied field. The relative intensity of the higher subband

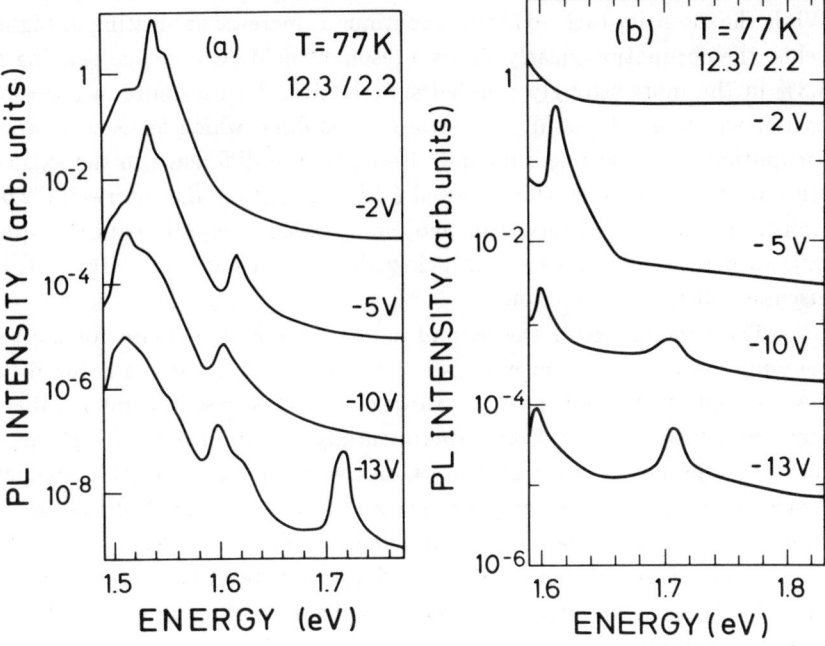

Fig. 4.9. Photoluminescence spectra 200 ps after pulsed photoexcitation for a superlattice with 12.3 nm GaAs wells, 2.2 nm AlAs barriers, and 50 periods for different applied voltages at 77 K. The spectra in (a) were recorded with a laser energy of 1.90 eV, while in (b) the excitation energy was reduced to 1.57 eV.

PL with regard to the band-gap PL can be deduced from Eq. 4.8

$$\frac{I_{i1}}{I_{11}} = \frac{O_{i1}}{O_{11}} \frac{n_i}{n_1} . \tag{4.10}$$

The ratio of the overlap integrals is equal to the ratio of oscillator strengths f_{ij}. The ratio of the overlap integrals or oscillator strengths in Eq. 4.10 can either be determined experimentally or by calculating the overlap integrals as described above. Since the oscillator strength ratio in Eq. 4.10 increases with increasing electric field, while a resonant behavior of the occupation ratio is expected, the relative intensity should increase with increasing field, but not necessarily exhibit a resonant behavior. This is demonstrated in Fig. 4.10, where a comparison of the field dependence of the experimentally measured intensity ratio with the occupation ratio is shown. The oscillator strength ratio was determined from field-dependent photocurrent spectra.[40] While the intensity ratio exhibits a continuous increase saturating at higher fields, the occupation clearly shows a resonant field dependence peaking at 3.3% in the more strongly coupled superlattice. Furthermore, a narrower barrier width should result in a stronger coupling, which leads to a larger occupation of the second subband. Recently, the difference in the exciton binding energy between the intrawell $C1H1$ transition and interwell $C2H1$ transition has been observed in photoluminescence experiments.[44] Superlattices with a subband spacing below the optical phonon energy will be discussed in the next section.

The occupation of the second subband should increase for a fixed coupling between adjacent wells, when the subband spacing is reduced below the optical phonon energy. However, in this case the intersubband scattering time has not been experimentally determined to a high accuracy. A superlattice with 21.0 nm GaAs wells and 2.2 nm AlAs barriers clearly showed the expected PL spectra, but the PL signal of the second subband was much smaller than anticipated. Furthermore, the oscillator strengths of the $C2H1$ and $C3H1$ could not be determined experimentally using photocurrent spectroscopy due to a strong mixing of heavy-hole and light-hole subbands.

4.3.3.3. *Electroluminescence spectroscopy*

In addition to infrared experiments there is another technique to investigate the electrical injection of carriers into higher subbands without photoexcitation. When the p-i-n diode is biased in the forward direction, electrons and holes are injected from the n- and p-contact, respectively. Since the superlattice forms the intrinsic region, the injected electrons and holes can recombine in the superlattice leading to the emission of electro-luminescence. This is actually a typical device configuration for light

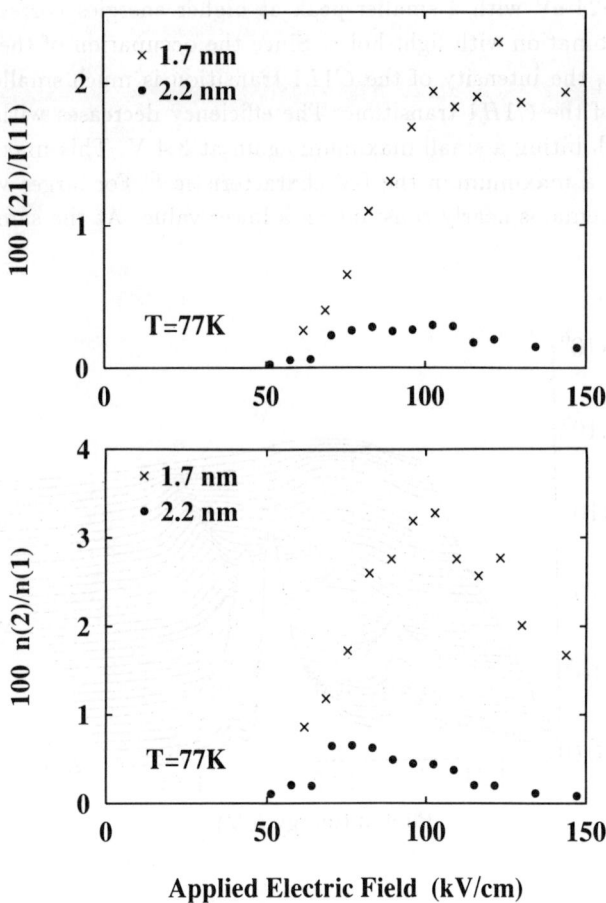

Fig. 4.10. Photoluminescence intensity (top) and occupation ratio (bottom) in two superlattices with similar GaAs well width (11.8 and 12.3 nm), but different AlAs barrier width as indicated at 77 K. The field strengths have been calculated using Eq. 4.6.

emitting diodes. The I-V characteristic of such a structure exhibits several resonances, which can be ascribed to resonant electron tunneling from the ground state ($C1$) into the second and third subband. The electroluminescence (EL) spectra for a superlattice with 21.0 nm GaAs wells, 2.2 nm AlAs barriers, and 40 periods is shown in Fig. 4.11. The electroluminescence intensity has been divided by the current, resulting in the EL efficiency. The spectrum is dominated at lower voltages by a strong emission line at 1.524 eV with a smaller peak at higher energies corresponding to the recombination with light-holes. Since the occupation of the light-holes is thermal, the intensity of the $C1L1$ transition is much smaller than the intensity of the $C1H1$ transition. The efficiency decreases with increasing voltage exhibiting a small maximum again at 3.4 V. This maximum coincides with a maximum in the I-V characteristic.[43] For larger voltages the efficiency remains nearly constant on a lower value. At the same time, the

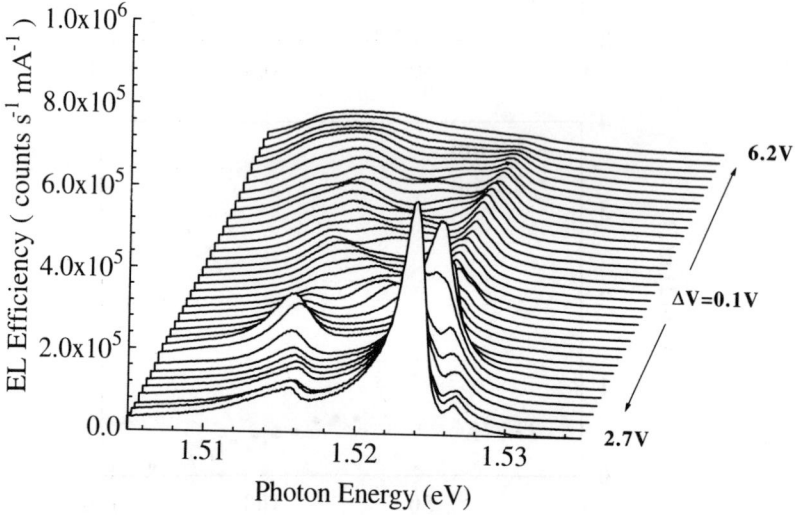

Fig. 4.11. Electroluminescence efficiency (intensity normalized to the current) vs photon energy in a superlattice with 21.0 nm GaAs wells, 2.2 nm AlAs barriers, and 40 periods at 5 K for applied voltages between 2.7 and 6.2 V in 0.1 V steps. Only the vicinity of the band-gap transition $C1H1$ is shown (from Ref. 43).

EL line broadens strongly and splits up into several peaks at lower photon energies. These peaks shift to red with increasing voltage.

In Fig. 4.12 the emission spectrun of the same sample is shown between 20 and 100 meV above the $C1H1$ transition. At 2.8 V a new line emerges at 1.558 eV exhibiting an efficiency maximum at 2.4 V. This line is identified as the $C2H1$ transition. The line width is rather narrow compared to the line width of the band-gap transition in Fig. 4.11. The $C2H1$ emission line is accompanied by two weaker EL lines, one at lower energies (1.546 eV) and one at higher energies (1.563 eV). The latter is clearly the $C2L1$ transition. However, the former one can only by assigned to the $C1H3$ transition and originates therefore from resonant tunneling between heavy-hole subbands. For voltages above 3.6 V the efficiency of the $C2H1$ transition remains constant again. At the same time the peak position does

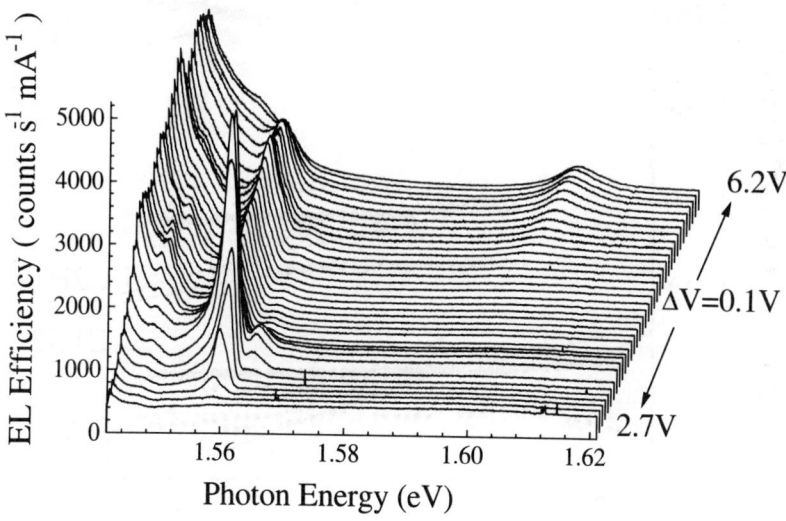

Fig. 4.12. Electroluminescence efficiency (intensity normalized to the current) vs photon energy between 20 and 100 meV above the band-gap transition $C1H1$ in a superlattice with 21.0 nm GaAs wells, 2.2 nm AlAs barriers, and 40 periods at 5 K for applied voltages between 2.7 and 6.2 V in 0.1 V steps. The efficiency is about 200 times less than that of the $C1H1$ transition in Fig. 4.11 (from Ref. 43).

not shift with increasing voltage as is expected from the QCSE. Above 6 V the efficiency suddenly decreases and the line shifts to lower energies. At the same time a new EL line appears at 1.60 eV, which corresponds to the $C3H1$ transition. This line is much broader than the $C2H1$ transition. The efficiency of the higher subband EL lines is about 0.5% of the band-gap EL efficiency.

In Fig. 4.13 the experimentally observed positions of the EL lines are compared to the calculated transition energies as a function of the applied voltage. The EL peak positions derived from Fig. 4.11 and 4.12 are shown as dots. The large dots mark the energetic positions of the dominating emission lines. The calculated transition energies for the interband emission involving the indicated subbands are shown as solid lines. The energies of the $CiHj$ transition are calculated from the respective subband energies

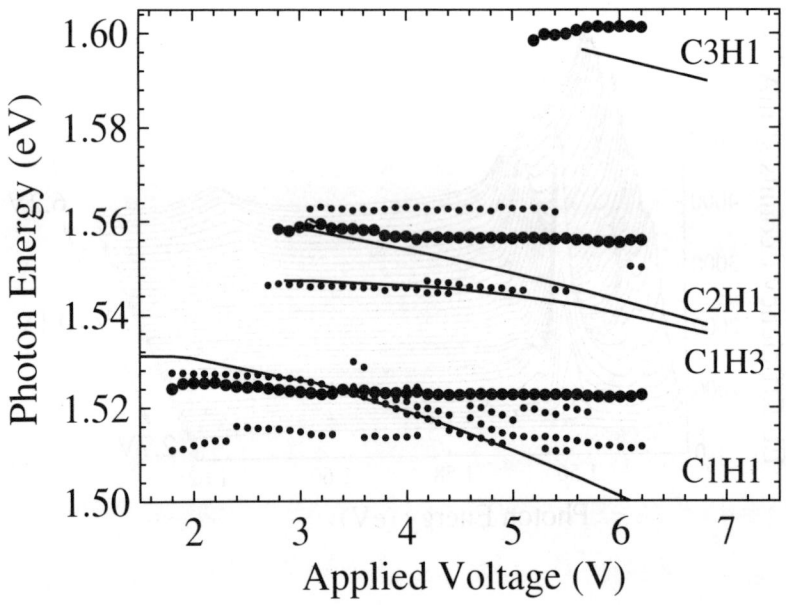

Fig. 4.13. Electroluminescence energy peak positions of Figs. 4.11 and 4.12 vs applied voltage (dots). The large dots mark the $C1H1$, $C2H1$, and $C3H1$ emission lines (from bottom to top). The solid lines indicate the calculated interband transition energies (from Ref. 43).

energies plus the energy gap of GaAs at low temperatures (1.52 eV). To account for the exciton binding energy, all calculated transition energies are shifted by a few meV so that the calculated $C1H1$ transition energy coincides with the experimental value derived from photoluminescence experiments at flat-band. The difference in exciton binding energies between different subbands is neglected as well as the shift of the exciton binding energy with electric field. The $C2H1$ ($C3H1$) transition appears at the calculated resonance voltage for tunneling from $C1$ into $C2$ ($C3$). Transitions between higher subbands in conduction and valence band are not observed, because their signal would be again two orders of magnitude smaller.

It is rather surprising to observe that the $C1H1$ and $C2H1$ transitions do not exhibit the expected Stark-shift with increasing applied voltage, but remain at the same energetic position. The emission peaks of these transitions must originate from regions of nearly constant field within the superlattice despite of the increasing applied voltage. The injected carriers are partially screening the applied electric field. We will return to this subject in Chapter 5, where the formation of electric-field domains is presented.

The intensity ratio of the $C2H1$ and $C1H1$ transition at the voltage of 3.4 V is determined to be 0.011. The ratio of the oscillator strengths can be estimated from the calculation of the overlap integrals, $f_{21}/f_{11} = 0.93$ (cf. Fig. 4.8). The resulting ratio n_2/n_1 is 0.012, but the actual occupation is larger, since only part of the superlattice is in resonance (due to the field screening) and reabsorption of the above band-gap light reduces the observed intensity. In conclusion of this section, the relative occupation of the second subband did not increase significantly for a subband spacing below the optical phonon energy. Two possible reasons are considered. First, the injection in some wells is hindered by the slow intersubband relaxation time so that only a fraction of the wells are in resonance. Second, the intersubband relaxation time is shorter than assumed. Recent measurements using time-resolved infrared spectrosopy indicated an intersubband relaxation time as short as 40 ps.[45] Since the resonant transport times are also in this time range, this could strongly reduce the occupation of the second subband.

4.3.4. *Quantum cascade laser*

In order to produce an intersubband laser, the intersubband relaxation time has to be longer than the tunneling time into the inverted state. This was achieved by Faist et al.[11] using a triple well structure separated by very thin barriers as the building block of the superlattice. Furthermore, the injector between the triple well structures was designed as a digitally graded alloy in order to achieve the flat-band condition in this layer at resonance. Two periods of this structure are shown schematically in Fig. 4.14. The triple well system contains three $In_{0.53}Ga_{0.47}As$ wells of thickness 0.8, 3.5, and 2.8 nm separated by $In_{0.52}Al_{0.48}As$ barriers of thickness 3.5 and 3.0 nm. The barrier at the injector is 4.5 nm thick, while the barrier at the collector is 3.0 nm wide. The calculated differences between the three levels in the triple well system are $E_{C3} - E_{C2} = 295$ meV and $E_{C2} - E_{C1} = 30$ meV.

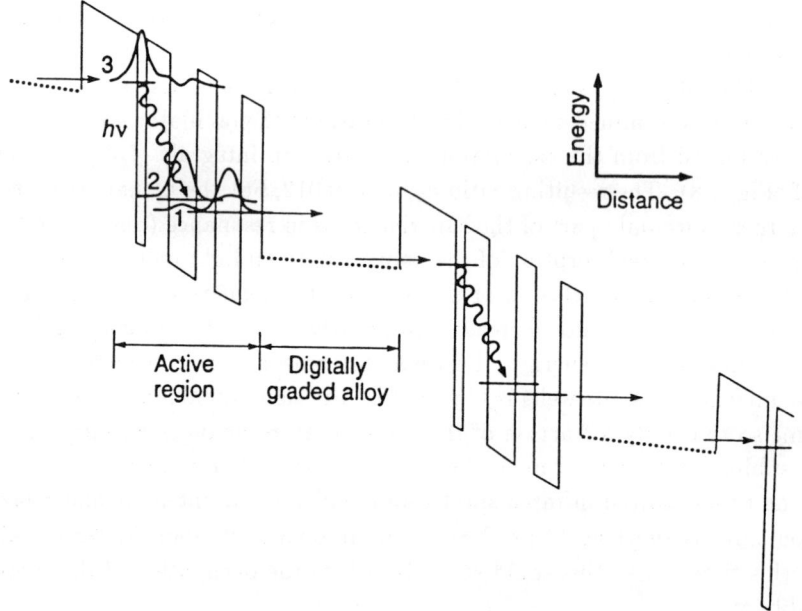

Fig. 4.14. Conduction band energy diagram of a portion of the 25-period section of the quantum cascade laser. The dashed lines are the effective conduction band edges of the digitally graded electron-injecting regions (from Ref. 11). The details of the structure are discussed in the text.

The structure was designed in such a way that the relaxation time between levels 3 and 2 is about a factor of 20 larger than the tunneling time into level 3. At the same time, the relaxation from level 2 to 1 was almost an order of magnitude faster than from level 3 to level 2. Finally, the tunneling escape time out of level 1 into the graded region was again even faster than the relaxation from level 2 to 1. The expected laser transition is between levels 3 and 2, i.e., at 295 meV or 4.2 μm. This spacing is well above the optical phonon energy. Only by using spatially indirect transitions, the occupation could be inverted. In Fig. 4.15 the emission spectrum of the laser for different currents is shown at 10 K. The strong line narrowing and large increase in optical power above 850 mA clearly demonstrates lasing action. Since the discovery of the quantum cascade laser the operating temperature has increased to above 100 K. So far the

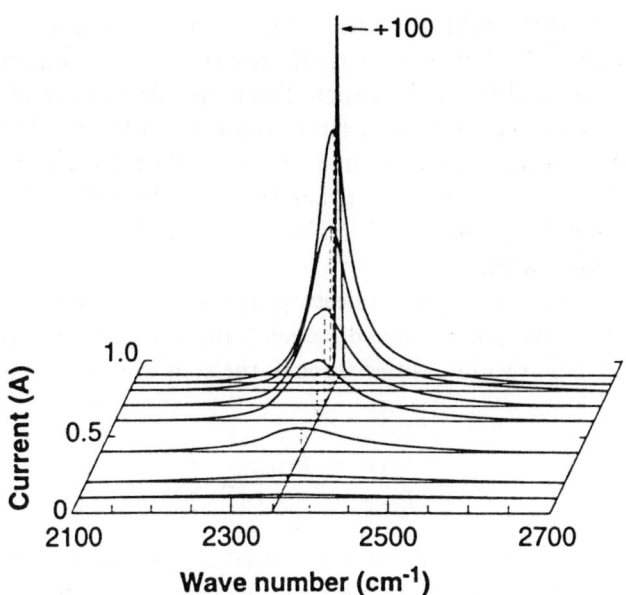

Fig. 4.15. Emission spectrum of the quantum cascade laser at various drive currents. The emission wavelength is 4.26 μm at 10 K (from Ref. 11).

laser operates under pulsed excitation. However, due to the intersubband nature of the lasing action, it is a unipolar device.

The quantum cascade laser is the realization of a 25-year-old idea by Kazarinov and Suris.[9] Hopefully, it will stimulate more work on intersubband lasers, also in the far-infrared regime, where intersubband relaxation by emission of optical phonons is forbidden.

4.3.5. *The quantum-well Pockels effect*

A rather peculiar effect was discovered by probing the excited subband photoluminescence in an applied electric field, the quantum-well Pockels effect.[46,47] In zinc blende semiconductors, the application of a [001] electric field leads to a biaxial behavior, where the anisotropy in the plane perpendicular to the field distinguishes [110] from [1$\bar{1}$0].[48] However, this Pockels effect remains very small up to the breakdown field[49] and cannot be observed in band-gap photoluminescence experiments. The photoluminescence from the second subband, which is forbidden at zero electric field, but becomes allowed as the field increases, exhibits a very strong anisotropy. In Fig. 4.16 typical PL spectra of the $C2H1$ transition are shown for different applied voltages and two polarization directions. At smaller voltages the anisotropy is much larger than at larger voltages. In the $C1H1$ PL signal no polarization dependence was found. Although the intensity of the $C2H1$ PL line is about 10^{-4} times weaker than the intensity of the $C1H1$ PL line in weakly coupled superlattice, the anisotropy is much more pronounced in the higher subband PL.

The degree of the in-plane anisotropy is measured by recording the PL intensity for the two polarization direction [110] and [1$\bar{1}$0]. By dividing the difference of these two intensities through the sum of them, a polarization ratio is determined, i.e.,

$$\rho = \frac{I_{[110]} - I_{[1\bar{1}0]}}{I_{[110]} + I_{[1\bar{1}0]}} = \frac{I_{\parallel}^{-}}{I_{\parallel}^{+}}. \tag{4.11}$$

In Fig. 4.17 this polarization ratio is plotted as a function of the applied voltage (or electric field). The $C2H1$ PL line cannot be detected for small electric fields. However, for electric fields above 30 kV cm^{-1}, the polarization ratio decreases in both field directions in the same way. Both, the large magnitude and the unusual *decrease* in anisotropy, are beyond an interpretation based solely on bulk properties. However, they can be

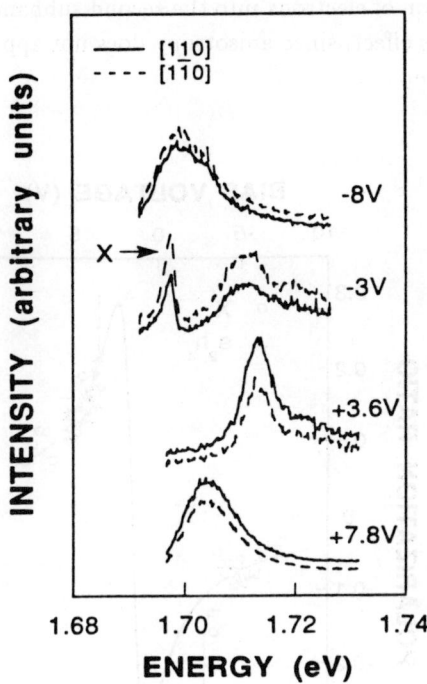

Fig. 4.16. Polarized photoluminescence spectra of the $C2H1$ transition in a superlattice with 9.0 nm GaAs wells, 4.0 nm AlAs barriers, and 40 periods for different applied voltages at 2 K. The peak labeled X is due to barrier $\rightarrow$ well recombination involving AlAs X-electron states. The excitation energy is 1.833 eV with a power density of 4 W cm^{-2} (from Ref. 45).

accounted for by quantum-well effects. Beyond linear response the data can be described by

$$\rho = \frac{F/F_1}{1 + (F/F_2)^2} , \tag{4.12}$$

where F_1 and F_2 are two parameters, which are used in the least-squares fit also shown in Fig. 4.17. The obtained values agree resonably well with estimates based on a model, which takes into account the mixing of the heavy- and light-hole subbands for F_1 and the behavior of the envelope wave function for F_2. This effect was observed in another weakly coupled superlattice with 8.0 nm GaAs, 11.0 nm AlAs barriers, and 10 periods. In more strongly coupled superlattices it has not been observed so far. The continuous injection of electrons into the second subband is essential for the observation of this effect, since anisotropy does not appear in the band-gap photoluminescence.

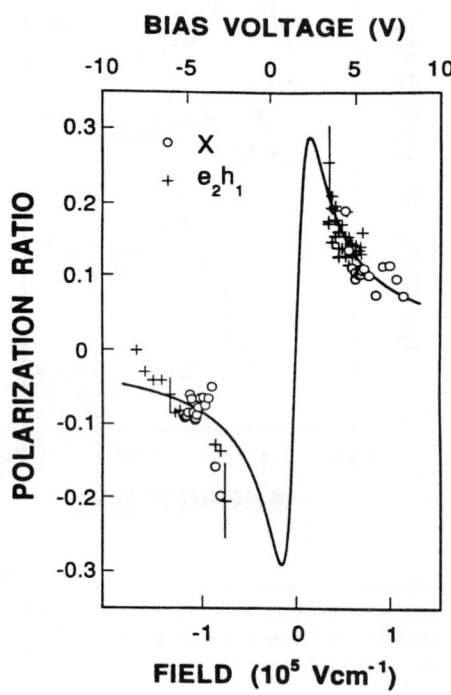

Fig. 4.17. Polarization ratio ρ for the $XH1$ and $C2H1$ transitions as a function of the perpendicular electric field and bias voltage for a superlattice with 9.0 nm GaAs wells, 4.0 nm AlAs barriers, and 40 periods at 2 K. The solid line is a least-squares fit to the data points. The electric field has been calculated using Eq. 4.6 with a width of the intrinsic region of 0.57 μm (from Ref. 45).

4.4. Magnetotunneling of electrons

The application of a magnetic field perpendicular and parallel to the layers in addition to the electric field can strongly influence the resonant tunneling process. In the perpendicular configuration the motion due to the electric and magnetic fields decouples. A complete quantization into Landau-levels occurs, i.e., the system behaves quasi zero-dimensional. For a parallel magnetic field the resonant tunneling process is strongly affected, since a deflection of the electrons in the superlattice direction takes place due to the magnetic field. Furthermore, the resonance condition is also influenced by the magnetic field. In both cases, magentotunneling between strongly confined, identical two-dimensional systems can be investigated in contrast to double-barrier systems,[50-59] in which the influence of the difference between the emitter and the quantum well has to be taken into account.

4.4.1. *Landau-level tunneling*

The effect of the magnetic field perpendicular to the layers, i.e., parallel to the electric field, leads to a quantization of the two-dimensional subbands into Landau-levels. Since the effect of the electric and magnetic field decouple in the Schrödinger equation, it can be solved in a similar manner as before. Neglecting spin effects, the Hamiltonian is given by

$$H = \frac{1}{2\,m_\parallel^*}\,(\vec{p}_\parallel - e\,\vec{A})^2 + \frac{p_z^2}{2\,m_z^*} + V(z)\,, \qquad (4.13)$$

where $V(z)$ includes the electric-field dependent quantum well potential. $\vec{A}$ denotes the vector potential, which is in the Landau gauge related to the magnetic field through $\vec{A} = B(0, x, 0)$. The solution of the Schrödinger equation is given by

$$E_{i,n} = E_i + \left(n + \frac{1}{2}\right)\hbar\omega_c\,, \qquad (4.14)$$

where E_i denotes the subband energy of the ith subband, $\omega_c = eB/m_\parallel^*$, $m_\parallel^*$ the in-plane effective mass at the bottom of the subband, and $n = 0, 1, 2, 3, \ldots$ Assuming that only the lowest Landau-level in the injecting well is occupied, the electric field strengths for resonant tunneling between adjacent wells are determined by the energy difference

$$E_{1,n}(B) - E_{1,0}(B) = n\,\hbar\omega_c\,. \qquad (4.15)$$

However, resonant inter-Landau-level tunneling is forbidden, since the in-plane wave functions in a magnetic field are harmonic oscillator eigenfunctions, which are orthogonal. This selection rule holds in the absence of any

scattering processes. Elastic scattering processes such as interface roughness or impurity scattering as well as inelastic scattering by emission of an
optical phonon can lead to violation of the selection rule and the observation of inter-Landau-level tunneling resonances.[60-62] Furthermore, the
non-parabolicity of the conduction band effective mass leads to a modification of the condition in Eq. 4.15. Following Ekenberg,[63], who calculated
energy spectrum of a quantum well in a perpendicular magnetic field, the
effective mass in Eq. 4.15 has to be replaced by

$$\frac{1}{m_{\parallel}^*} - \frac{[(8n^2 + 8n)\,\alpha' + (n^2 + n)\,\beta']}{8\,n} \frac{\hbar\,e\,B}{m_1^{*2}}\,, \qquad (4.16)$$

where $\alpha' = 0.642$ eV^{-1} and $\beta' = 0.697$ eV^{-1} are the non-parabolicity
parameters of GaAs[63,64] and m_1^* is the effective mass of bulk GaAs at the
conduction band edge, i.e., $m_1^* = 0.0665\ m_0$.

Resonant tunneling between Landau levels was observed in a reverse
biased p-i-n structure with a very small photoexcited carrier density of less
than 10^9 cm^{-2}. In Fig. 4.18 two photocurrent-voltage characteristics are
shown, one at vanishing magnetic field, the other at 8 T. The trace at 0 T
exhibits several resonances which can be ascribed to resonant tunneling
between the first and second electronic subband (-4.7 V) and between the
first heavy- and light-hole subbands (0.75 V). The other two resonances are
due to phonon-assisted tunneling between the ground levels (-0.1 V) and
the first heavy- and light-hole subbands (-0.8 V). The trace at 8 T, however, shows a larger number of oscillations, which can be better resolved
by subtracting the 0-T trace from it (cf. inset of Fig. 4.18). The period
of the oscillations corresponds to an energy difference of about 12.5 meV,
which approximately agrees with the cyclotron energy $\hbar\omega_c$ using a slightly
heavier electron effective mass than in bulk GaAs. Therefore, these peaks
can be identified as inter-Landau-level resonances of electrons between adjacent wells. The calculated subband spacing between the first and second
electronic subband is 135 meV so that even at 8 T at least nine Landau
levels fit between these subbands.

In Fig. 4.18 resonances are observed below and above the optical phonon energy, i.e., the inter-Landau-level tunneling process is accompanied
by elastic as well as inelastic scattering. Focusing first on the inelastic tunneling resonances, the energy positions ($E = edF_{ap}$) of the resonance peaks
above 30 meV are plotted versus the applied magnetic field. The size of

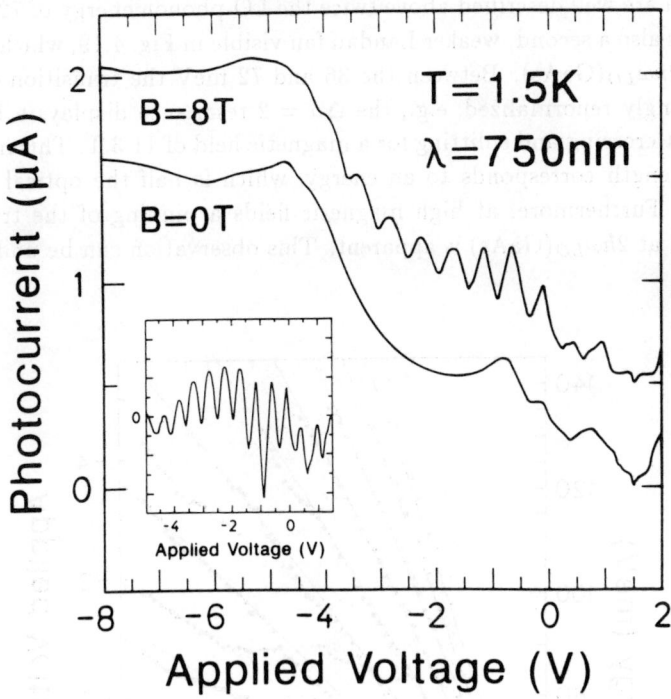

Fig. 4.18. Absolute value of the photocurrent at 0 and 8 T vs applied voltage in a superlattice with 9.0 nm GaAs wells, 4.0 nm AlAs barriers, and 40 periods at 1.5 K. The trace at 8 T has been shifted by 0.5 μA. The inset shows the difference between the 8-T and 0-T trace (from Ref. 60).

the dots is proportional to the relative strength of the peaks in the I-V characteristic. A Landau fan converging for zero magnetic field to the GaAs LO-phonon-assisted resonance at 36.2 meV is clearly resolved. Tunneling resonances with a change of the Landau index Δn between 0 and 9 are visible. Due to the low carrier concentration, electrons are only injected from the $n = 0$ Landau level into higher levels in the adjacent well. Because of the large barrier height and the weak coupling between adjacent wells, the injected electrons relax to the ground level before tunneling into the next well can occur. The solid lines are a fit to the data points using the expression in Eq. 4.15 with the non-parabolicity corrections from Eq. 4.16. The only adjustable parameter is the parallel mass $m^*_\parallel$. Using $m^*_\parallel = 0.069\ m_0$, the data are well described above twice the LO-phonon energy of 72.4 meV. There is also a second, weaker Landau fan visible in Fig. 4.19, which converges to $2\hbar\omega_{LO}(\text{GaAs})$. Between the 36 and 72 meV the transition energies are strongly renormalized, e.g., the $\Delta n = 2$ resonance displays a Landau-level anticrossing and splitting for a magnetic field of 11.3 T. This magnetic field strength corresponds to an energy, which is half the optical phonon energy. Furthermore, at high magnetic fields a pinning of the transition energies at $2\hbar\omega_{LO}(\text{GaAs})$ is apparent. This observation can be understood

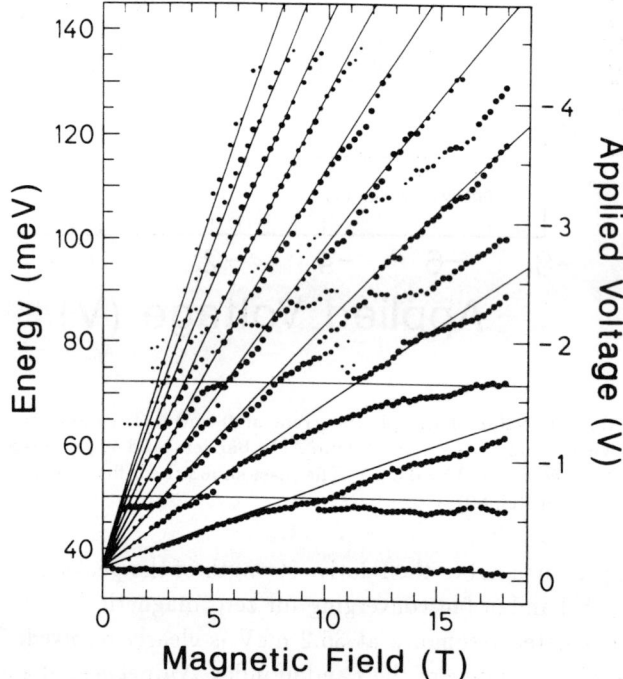

Fig. 4.19. Peak positions for the LO-phonon-assisted tunneling inter-Landau-level tunneling (dots) and corresponding Landau fan for a superlattice with 9.0 nm GaAs wells, 4.0 nm AlAs barriers, and 40 periods taking conduction non-parabolicities into account (solid lines). The size of the dots indicates the relative strength of the peaks. The energies of the GaAs LO phonon (36.2 meV), the AlAs LO phonon (50.1 meV), and twice the GaAs LO phonon (72.4 meV) are marked by horizontal lines (from Ref. 60).

within the model of 2-dimensional magnetopolarons.[65-67] Due to the interaction of the electrons with LO phonons in the GaAs wells, which is usually referred to as the polaron, the Landau-level energies are shifted by an amount $\Delta E_n(B)$. For weak electron-phonon coupling the energy shift can be calculated in second order perturbation theory resulting in an implicit equation for $\Delta E_n(B)$.[61] Taking into account the non-parabolicity of the effective mass[63] and a polaron coupling constant of 0.07 for GaAs,[67] the calculated Landau level resonances shown in Fig. 4.20 are obtained. No fitting parameters were used. In particular, for $\Delta n = 1$ and 2 the data agree well with the calculated curves.

The horizontal data points in Fig. 4.20 at about 48 meV are assigned to scattering with AlAs-like interface phonons.[68-70] Since these interface phonons also produce some long-range electric field in the GaAs wells,[68] their coupling to the tunneling electrons might be stronger than for the AlAs slab phonons. The calculated energies of the two AlAs-like interface modes for small phonon wave vector, 46.5 and 48.5 meV, are rather close

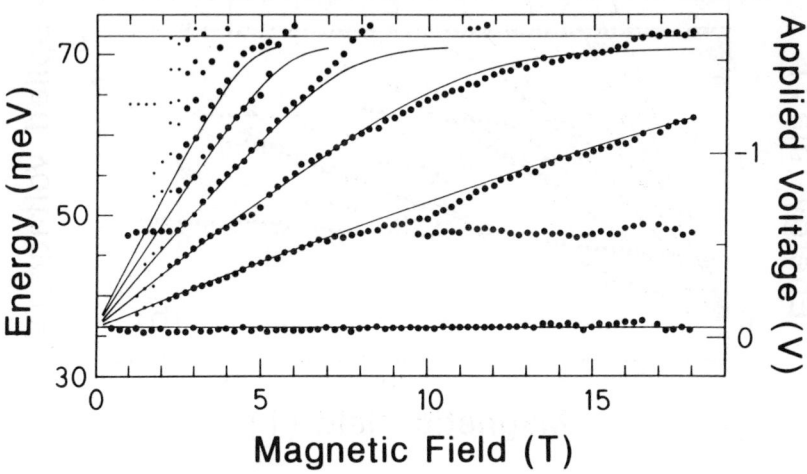

Fig. 4.20. Peak positions for LO-phonon-assisted inter-Landau-level tunneling (dots) between $\hbar\omega_{LO}$(GaAs) and $2\hbar\omega_{LO}$(GaAs) as well as calculated tunneling resonance energies including 2D magnetopolaron corrections and non-parabolicities for $1 \leq \Delta n \leq 5$ (solid lines) for a superlattice with 9.0 nm GaAs wells, 4.0 nm AlAs barriers, and 40 periods at 1.5 K (from Ref. 60).

to the observed value.

The observation of resonant inter-Landau-level tunneling with the emission of optical phonons is not surprising. However, the Landau-level resonances below the optical phonon energy are rather unexpected. In

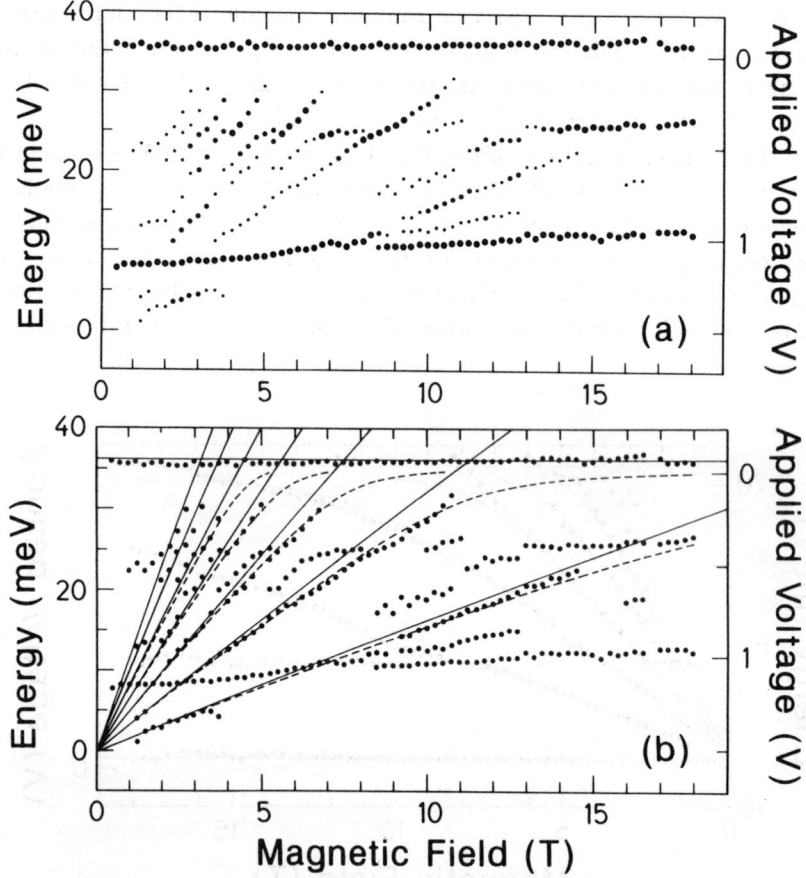

Fig. 4.21. (a) Peak positions of inter-Landau-level resonances below the GaAs LO-phonon emission line in a superlattice with 9.0 nm GaAs wells, 4.0 nm AlAs barriers, and 40 periods at 1.5 K. The size of the dots indicates the relative strength of the peaks. (b) Transition energies due to elastic scattering (dots) and calculated Landau fan without (solid lines) and with the inclusion of magnetopolaron corrections (dashed lines) (from Ref. 60).

Fig. 4.21(a) the observed inter-Landau-level resonances for energies below the GaAs LO-phonon emission line are shown. The elastic scattering resonances are possible because of symmetry-breaking fluctuations in barrier thickness, impurity scattering, or scattering due to interface defects. In Fig. 4.21 the corresponding Landau fan is plotted without (solid lines) and with the inclusion of magnetopolaron corrections (dashed lines). The nonparabolicity was included in the calculation in the same way as described above. Again, the agreement between experimental data and the calculation is improved, when the 2D magnetopolaron effects are taken into account.

There are also two resonances in Fig. 4.21 between 8 and 12.5 meV (25 and 26 meV), which exhibit only a weak magnetic field dependence. These rather strong resonances can be qualitatively explained by zone-folded acoustical phonons. Since the acoustical branches of the constituent materials in a superlattice always overlap, the new periodicity of the superlattice leads to new *optical branches* in the acoustical phonon energy regime.[70,71] The observed resonance positions correspond to phonon energies with a peak in the density of states, i.e., the zone-folded TA and LA modes of the bulk zone-boundary phonons.

A second superlattice with a wider well width, i.e., smaller subband spacing, but similar coupling, was investigated in order to study the interaction of Landau level originating from different subbands. However, the observed resonances could not be explained by electron tunneling. The results will be discussed in section 4.5 in connection with resonant hole tunneling.

4.4.2. *Tunneling in crossed electric and magnetic fields*

In crossed electric and magnetic fields the tunneling resonances are shifted to higher electric field strengths. This can be qualitatively explained using semiclassical arguments. Choosing the magnetic field direction parallel to the x-axis, a carrier that travels the *tunneling distance d* from one well to the next picks up momentum $\hbar k_y = eBd$ due to the Lorentz force. In the effective mass approximation this corresponds to an energy $\Delta E = e^2 B^2 d^2 / (2m^*)$ that is transferred from the z to the y direction. Therefore, in order to maintain the resonance condition, this energy transfer has to be compensated by an additional electric field $\Delta F = \Delta E/(ed)$. These rather crude arguments can be quantified using perturbation theory for the magnetic field. In the effective mass approximation and Landau gauge the energies are given by[72]

$$E_i(B) = E_i(0) + \frac{\hbar^2 k_x^2}{2 m_i^*} + \frac{e^2 B^2}{2 m_i^*}(\langle z^2 \rangle_i - \langle z \rangle_i^2)$$
$$+ \frac{1}{2 m_i^*} (\hbar k_y + eB \langle z \rangle_i)^2 ,$$

(4.17)

where m_i^* denotes the parallel mass at the bottom of the ith subband. The third term in this equation represents the diamagnetic shift, while the last term reflects the coupling of of momentum $\hbar k_y$ to the motion in the z direction due to the Lorentz force. Thus the parallel magnetic field leads to a parabolic energy dispersion perpendicular to the layers with a minimum for each well and subband being characterized by the constraint

$$k_y \mid_{min} = -\frac{eB}{\hbar} \langle z \rangle_i .$$

(4.18)

The resonant tunneling process between the first and second electronic subbands of adjacent wells in crossed electric and magnetic fields is illustrated in Fig. 4.22. A carrier that has relaxed to the minimum of the first electronic subband in the right well can resonantly tunnel into the left well when the parabolic dispersion of the second subband intersects this minimum. This occurs when an additional electric field ΔF is applied with respect to the zero magnetic field case. The corresponding energy shift $\Delta E = \Delta F e d$ is easily obtained using Eqs. 4.16 and 4.17[73]

$$\Delta E = \frac{e^2 B^2}{2 m_2^*} (\langle z_r \rangle_1 - \langle z_l \rangle_2)^2$$
$$+ \frac{e^2 B^2}{2 m_2^*} (\langle z_l^2 \rangle_2 - \langle z_l \rangle_2^2)$$
$$- \frac{e^2 B^2}{2 m_1^*} (\langle z_r^2 \rangle_1 - \langle z_r \rangle_1^2) ,$$

(4.19)

where r and l denote the left and right wells in Fig. 4.22, respectively. The first term in Eq. 4.18 originates from the momentum transfer due to the Lorentz force, whereas the last two terms reflect the difference in the diamagnetic shifts of the involved subbands. We note that the expectation values in Eq. 4.18 depend on the electric field strength due to the QCSE.[14] The subsequent relaxation process now takes place not only between subbands but also involves a momentum relaxation in order to fulfill Eq. 4.17 in the left well.

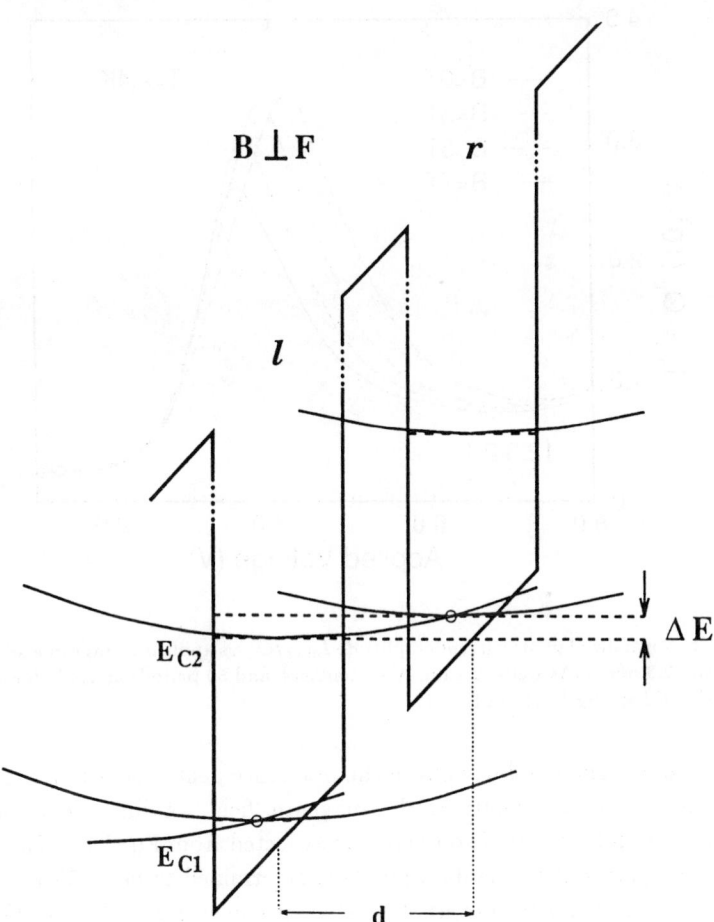

Fig. 4.22. Energy levels in a superlattice in crossed electric and magnetic fields. The first subband of the right well (r) is at resonance with the second subband ot the left well (l). The shift of the resonance ΔE with respect to $B = 0$ T is not drawn to scale.

The shift of the tunneling resonances is investigated best employing the time-of-flight technique. In Fig. 4.23 the normalized photocurrent amplitude I_{max}/Q_0 is plotted as a function of the applied voltage for a superlattice with 12.3 nm GaAs wells, 2.1 nm AlAs barriers, and 50 periods at 2.4 K. The resonances clearly shift to larger reverse bias voltages, i.e., larger

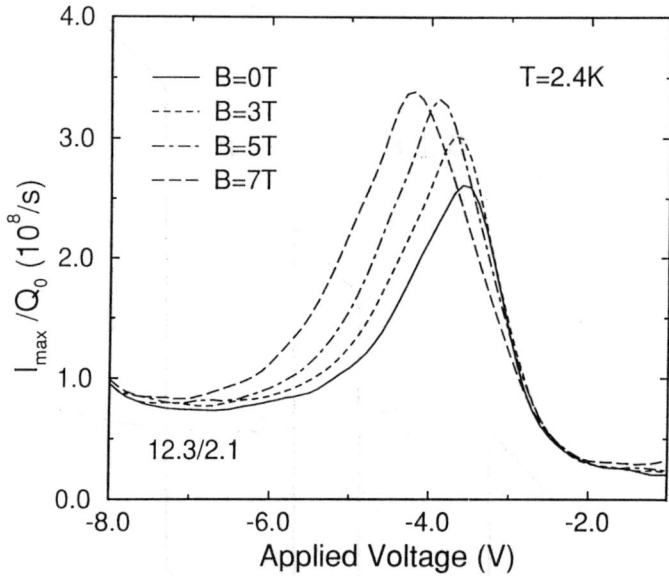

Fig. 4.23. Normalized photocurrent amplitude I_{max}/Q_0 vs applied voltage in a superlattice with 12.3 nm GaAs wells, 2.1 nm AlAs barriers, and 50 periods at 2.4 K for several magnetic field strengths (from Ref. 72).

electric fields. The absolute shift of the resonance peak is shown in Fig. 4.24 as a function of the square of the magentic field. A linear relationship between the shift and B^2 is observed as expected from Eq. 4.19. The slope of a least squares fit to the data points is determined to be $(1.55 \pm 0.05) \times 10^{-2}$ VT^{-2}. In order to convert this value into an energy shift, the effective width of the intrinsic region has to be determined. Since the resonance appears at -3.6 V for vanishing magnetic field, the subband spacing is 86 meV, and the built-in voltage 1.5 V, the effective width of the intrinsic region is $59.3d$ according to Eq. 4.6. The voltage shift therefore corresponds to an energy shift of $\Delta/B^2 = (2.60 \pm 0.09) \times 10^{-4}$ eVT^{-2}. Assuming the same mass for the first and second subband, the energy shift in Eq. 4.19 is determined by numerical calculation of the expectation values. For the investigated superlattice the result is $\Delta/B^2 = 2.38 \times 10^{-4}$ eVT^{-2} in very good agreement with the experimentally obtained value. A second sample with very different coupling has been investigated resulting in a similar

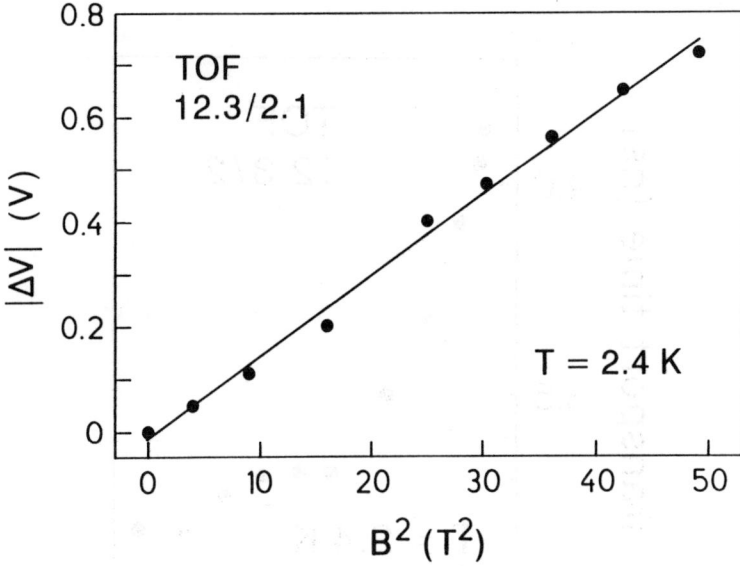

Fig. 4.24. Absolute shift ΔV of the resonance position vs the square of the magnetic field in a superlattice with 12.3 nm GaAs wells, 2.1 nm AlAs barriers, and 50 periods at 2.4 K (from Ref. 72).

agreement between the measured and calculated shift.[73]

    Finally, the dynamics of the resonant tunneling process in a parallel magnetic field will be briefly discussed. It is apparent from Fig. 4.23 that the peak value of the photocurrent amplitude increases with magnetic field. This is consistent with the observation that the corresponding transients become shorter. In Fig. 4.25 the transport times, i.e., the inverse of I_{max}/Q_0, are plotted as a function of the square of the magnetic field for a superlattice with 12.3 nm GaAs wells, 2.1 nm AlAs barriers, and 50 periods. This data set was obtained with a different mesa structure than the photocurrent amplitudes in Fig. 4.23. In a parallel magnetic field the resonance is shifted to higher electric fields. Therefore, the effective barrier height at resonance is reduced. This leads to an increase in the resonance splitting, and subsequently to a decrease in the resonant tunneling time. The observed saturation above 5 T is not fully understood, but may be related to the breakdown of the discussed resonant tunneling model at larger magnetic fields.

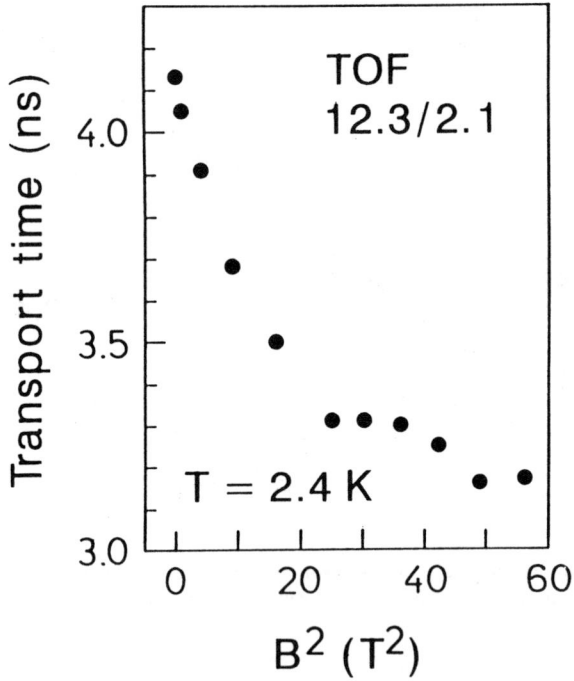

Fig. 4.25. Transport time through a superlattice with 12.3 nm GaAs wells, 2.1 nm AlAs barriers, and 50 periods at 2.4 K (from Ref. 72).

4.5. Resonant tunneling of holes

We have been so far only concerned with resonant electron tunneling. However, in forward biased p-i-n diodes, holes are also injected and contribute to the transport. In a reversed biased p-i-n diode carriers are introduced by photoexcitation, and, consequently, holes are also present. Hole resonances are usually difficult to detect, since they are masked by electron resonances, which move much faster at resonance than holes due to their smaller mass. It was demonstrated by Schneider et al.[32] that the time-of-flight transients are usually dominated by electron transport. In Fig. 4.26 transient photocurrents at a reverse bias of -2 V are shown for photoexcitation from the front and the back side in a superlattice with 12.3 nm GaAs wells, 2.1 nm barriers, and 50 periods. The p-contact is on the top of the sample, while the n-contact is achieved through the substrate. In order to perform the experiment from the back, the substrate had to be removed by etching. Since the transients with excitation from the back side are much shorter in time, the photocurrent is dominated by electron transport. In the backside illumination configuration, electrons and holes are excited near the n-contact, since the large photoexcitation energy limits the penetration depth to only about 20% of the intrinsic region of the p-i-n diode. The conclusion is that the hole transport occurs on a much longer time scale.

In order to observe resonant hole tunneling, the electronic subband spacing has to be increased so that no electronic tunneling resonances are present. This can be achieved by decreasing the GaAs well width down to 5 nm, while keeping a similar AlAs barrier width of 2.2 nm as before.

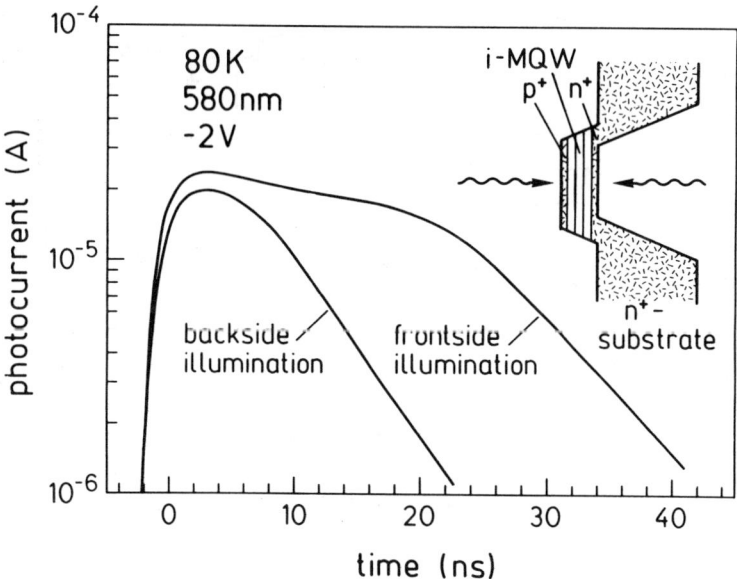

Fig. 4.26. Transient photocurrents at a reverse bias voltage of −2 V in a superlattice with 12.3 nm GaAs wells, 2.1 nm AlAs barriers, and 50 periods for frontside and backside illumination at 80 K with an excitation energy of 2.14 eV (from Ref. 32).

Resonant hole tunneling in superlattices was first observed by Schneider et al.[74] In Fig. 4.27 the peak photocurrent as a function of the applied voltage in a superlattice with 5.0 nm GaAs wells, 2.2 nm AlAs barriers, and 90 periods is shown for different temperatures. Several resonances are observed, which cannot be ascribed to resonant electron tunneling between adjacent wells, since the subband spacing between the first two electronic subband amounts to 340 meV resulting in a resonance voltage of about −30 V. Therefore, these resonances were attributed to resonant tunneling between the first heavy- and light-holes subband (-2.2 V) and between the first two heavy-holes subbands (−11 V). The resonance at low fields (0.5 V) could be identified as due to transport within the first heavy-hole miniband. The calculation of the hole subband spacings is much more involved due to the mixing of heavy- and light-holes states at finite wavevector. While

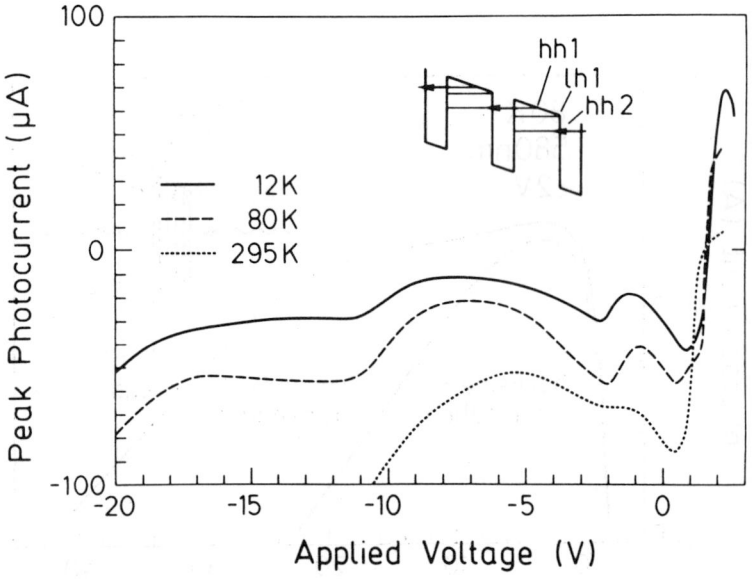

Fig. 4.27. Peak photocurrent vs applied voltage in a superlattice with 5.0 nm GaAs wells, 2.2 nm AlAs barriers, and 90 periods at different temperatures. The schematics of sequential resonant hole tunneling (corresponding to the situation at −11 V) is also shown (from Ref. 73).

there is no doubt about the identification of the resonance at -2.2 V, resonant-Raman-scattering experiments gave some evidence for another interpretation of the resonance at -11 V.[75] So far the X-levels in the AlAs barrier could be neglected, since the well widths used to study resonant electron tunneling are rather wide. However, an unambiguous identification cannot be achieved through a calculation of the separation between the $C1$-level and the $X1$-level in the barrier, since the corresponding band offsets are not too accurately known. The non-thermal occupation of higher heavy-hole subbands has been observed in the electroluminescence spectra of a GaAs-AlAs superlattice (cf. Fig. 4.12 and Ref. 43). There have been some investigations of resonant hole tunneling in double barrier[76] and

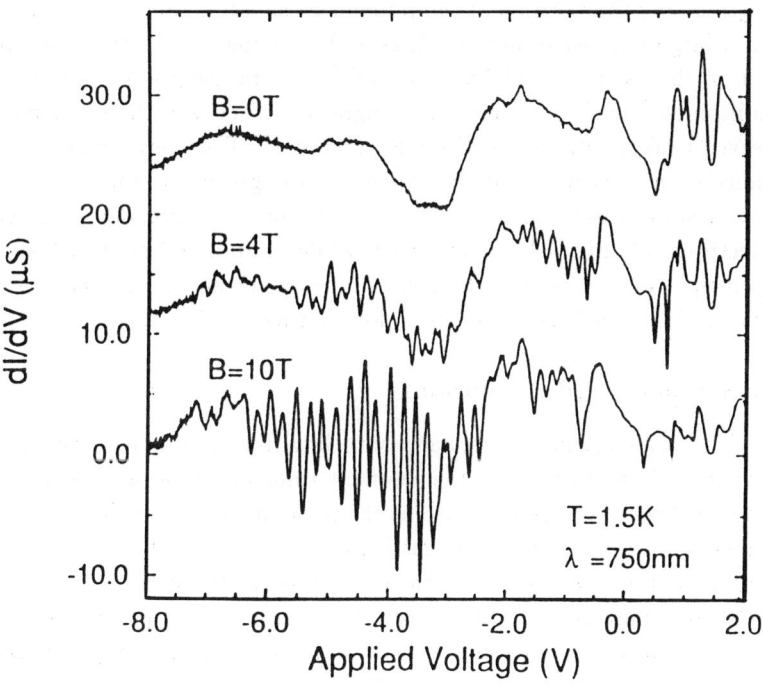

Fig. 4.28. Differential conductivity at perpendicular magnetic fields of 0, 4, and 10 T vs applied voltage in a superlattice with 14.4 nm GaAs wells, 3.4 nm AlAs barriers, and 50 periods. The traces at 0 and 4 T have been shifted by 12 and 24 μS, respectively (from Ref. 61).

asymmetric double quantum well structures.[77-79] Nevertheless, in contrast to resonant electron tunneling, there are still many open questions with regard to resonant hole tunneling.

Recently, resonant Landau-level tunneling of holes has been observed in GaAs-AlAs superlattices,[62] although the electronic subband spacing should also lead to tunneling resonances in the applied voltage range. In Fig. 4.28 the differential conductivity is shown for a superlattice in a p-i-n diode configuration with 14.4 nm GaAs wells, 3.4 nm AlAs barriers, and 50 periods. The p-i-n diode was biased in the reverse direction and illuminated with light of 1.653 eV to excite a very small carrier density below 10^9 cm^{-2}. The oscillations in the conductivity cannot be attributed to resonant Landau-level tunneling of electrons, since the spacing between peaks at electric fields beyond -2 V do not increase with magnetic field. Furthermore, plotting the position of the resonances as a function of the magnetic field, a large number of anticrossings and splittings is observed. A calculation of the valence band Landau-level structure using a 4×4 Luttinger Hamiltonian[80] leads to a qualitative agreement between the experimental observed peak positions and the calculated ones. However, further investigations are necessary to obtain a quantitative agreement. Finally, another open question arises, since in one sample resonant Landau-level tunneling of electrons (cf. Fig. 4.18) is observed, while in another one, which should in principle also show Landau-level tunneling of electrons, Landau-level tunneling between hole subbands is dominating (cf. Fig 4.28).

4.6. Summary and conclusions

The investigation of resonant tunneling in semiconductor superlattices yields information about the resonant tunneling process, but it is also useful for device applications such as the intersubband laser. The discovery of higher subband luminescence has led to the discovery of a new effect, the quantum-well Pockels effect, which is not observed in the band-gap luminescence. Resonant inter-Landau-level tunneling is a very useful method to study the non-parabolicity of the effective mass and the interaction between the electrons and longitudinal optical phonons, i.e., the magnetopolarons. Finally, the understanding of resonant hole tunneling without and with a perpendicular magnetic field is still not as advanced as the understanding of resonant electron tunneling. Further work is necessary to unambiguously indentify tunneling between heavy and light holes.

4.7. Acknowledgments

I would like to acknowledge the stimulating collaboration in this research area with D. Bertram, R. Klann, K. von Klitzing, S.H. Kwok, H. Lage, R. Merlin, W. Müller, K. Ploog, W.W. Rühle, and H. Schneider. Exceptional samples for some of this work were grown by A. Fischer.

References

1. F. Hund, *Z. Phys.* **40**, 742 (1927).
2. L. Nordheim, *Z. Phys.* **46**, 833 (1928).
3. R.W. Gurney and E.U. Condon, *Nature* **122**, 439 (1928).
4. G. Gamow, *Z. Phys.* **51**, 204 (1928).
5. C. Zener, *Proc. Roy. Soc. (London) A* **145**, 523 (1934).
6. L. Esaki, *Phys. Rev.* **109**, 603 (1958).
7. L.L Chang, E.E. Mendez, and C. Tejedor, *Resonant Tunneling in Semiconductors: Physics and Applications* (Plenum, New York, 1992).
8. L. Esaki and R. Tsu, *IBM J. Res. Develop.* **14**, 61 (1970).
9. R.F. Kazarinov and R.A. Suris, *Fiz. Tekh. Poluprov.* **5**, 797 (1971) [*Sov. Phys.-Semicond.* **5**, 707 (1971)].
10. H.T. Grahn, H. Schneider, W.W. Rühle, K. von Klitzing, and K. Ploog, *Phys. Rev. Lett.* **64**, 2426 (1990).
11. J. Faist, F. Capasso, D. Sivco, C. Sirtori, A.L. Hutchinson, and A.Y. Cho, *Science* **264**, 553 (1994).
12. C. Cohen-Tannoudji, B. Diu, and F. Laloë, *Quantum Mechanics Vol.1* (J. Wiley & Sons, New York 1977), p. 464.
13. S. Luryi, *Solid State Commun.* **65**, 787 (1988).
14. D.A.B. Miller, D.S. Chemla, T.C. Damen, A.C. Gossard, W. Wiegmann, T.H. Wood, and C.A. Burrus, *Phys. Rev. Lett.* **53**, 2173 (1984).
15. D.Y. Oberli, D.R. Wake, M.V. Klein, J. Klem, T. Henderson, and H. Morkoç, *Phys. Rev. Lett.* **59**, 696 (1987).
16. R. Ferreira and G. Bastard, *Phys. Rev. B* **40**, 1074 (1989).
17. J. Faist, C. Sirtori, F. Capasso, L. Pfeiffer, and K.W. West, *Appl. Phys. Lett.* **64**, 872 (1994).
18. K.T. Tsen and H. Morkoç, *Phys. Rev. B* **34**, 4412 (1986).
19. M.C. Tatham, J.F. Ryan, and C.T. Foxon, *Phys. Rev. Lett.* **63**, 1637 (1989).
20. B. Deveaud, F. Clerot, A. Chomette, A. Regreny, R. Ferreira, G. Bastard, and B. Sermage, *Europhys. Lett.* **11**, 367 (1990).
21. B. Deveaud, A. Chomette, F. Clerot, P. Auvray, A. Regreny, R. Ferreira, and G. Bastard, *Phys. Rev. B* **42**, 7021 (1990).
22. M.C. Tatham, A.M. De Paula, A.C. Maciel, and J.F. Ryan, *Semicond. Sci. Technol.* (1991).
23. D. Collings, K.L. Schumacher, F. Raksi, H.P. Hughes, and R.T. Phillips, *Appl. Phys. Lett.* **64**, 889 (1994).
24. L. Esaki and L.L. Chang, *Phys. Rev. Lett.* **33**, 495 (1974).
25. F. Capasso, K. Mohammed, and A.Y. Cho, *Appl. Phys. Lett.* **48**, 478 (1986).
26. T. Furuta, K. Hirakawa, I. Yoshino, and H. Sakaki, *Jpn. J. Appl. Phys.* **25**, L151 (1986).
27. S. Tarucha, K. Ploog, and K. von Klitzing, *Phys. Rev. B* **36**, 4558 (1987).
28. S. Tarucha, and K. Ploog, *Phys. Rev. B* **38**, 4198 (1988).
29. H. Schneider, K. von Klitzing, and K. Ploog, *Europhys. Lett.* **8**, 575 (1989).
30. H. Schneider, W.W. Rühle, K. von Klitzing, and K. Ploog, *Appl. Phys. Lett.* **54**, 2656 (1989).

31. H. Schneider, K. von Klitzing, and K. Ploog, *Superlattices Microstruct.* **5**, 383 (1990).

32. H. Schneider, H.T. Grahn, and K. von Klitzing, *Surf. Sci.* **228**, 362 (1990).

33. G. Livescu, A.M. Fox, D.A.B. Miller, T. Sizer, W.H. Knox, A.C. Gossard, and J.H. English, *Phys. Rev. Lett.* **63**, 438 (1989).

34. K. Leo, J. Shah, E.O. Göbel, T.C. Damen, S. Schmitt-Rink, W. Schfer, and K. Köhler, *Phys. Rev. Lett.* **66**, 201 (1991).

35. H. Roskos, M.C. Nuss, J. Shah, K. Leo, D.A.B. Miller, A.M. Fox, S. Schmitt-Rink, and K. Köhler, *Phys. Rev. Lett.* **68**, 2216 (1992).

36. W. Müller, D. Bertram, H.T. Grahn, K. von Klitzing, and K. Ploog, *Phys. Rev. B* **50**, 10998 (1994).

37. M. Helm, P. England, E. Colas, F. de Rosa, and S.J. Allen, *Phys. Rev. Lett.* **63**, 74 (1989).

38. M. Helm, in *Intersubband Transitions in Quantum Wells*, edited by E. Rosencher et al. (Plenum Press, New York, 1992), p. 151.

39. H.T. Grahn, W.W. Rühle, K. von Klitzing, and K. Ploog, *Semicond. Sci. Technol.* **7**, B409 (1992).

40. H.T. Grahn, W.W. Rühle, and K. Ploog, in *Quantum Well and Superlattice Physics IV*, edited by G.H. Döhler and E.S. Koteles, (SPIE, Bellingham, 1992), p. 36.

41. D. Bertram, H. Lage, H.T. Grahn, and K. Ploog, *Appl. Phys. Lett.* **64**, 1012 (1994).

42. H.T. Grahn, D. Bertram, H. Lage, K. von KLitzing, and K. Ploog, *Semicond. Sci. Techn.* **9**, 537 (1994).

43. R. Klann, H.T. Grahn, and K. Ploog, *Phys. Rev. B* **50**, 11037 (1994).

44. H. Schneider, J. Wagner, and K. Ploog, *Phys. Rev. B* **48**, 11051 (1993).

45. B.N. Murdin, G.M.H. Knippels, A.F.G. van der Meer, C.R. Pidgeon, C.J.G.M. Langerak, M. Helm, W. Heiss, K. Unterrainer, E. Gornik, K.K. Geerinck, N.J. Hovenier, and W.Th. Wenckebach, *Semicond. Sci. Technol.* **9**, 1554 (1994).

46. S.H. Kwok, H.T. Grahn, K. Ploog, and R. Merlin, *Phys. Rev. Lett.* **69**, 973 (1992).

47. S.H. Kwok, H.T. Grahn, K. Ploog, and R. Merlin, in *Proceedings of the 21nd International Conference on the Physics of Semiconductors*, edited by P. Jiang and H.-Z. Zheng (World Scientific, Singapore, 1992), p. 1180.

48. See, e.g., J.F. Nye, *Physical Properties of Crystals* (Clarendon Press, Oxford, 1985), Chap. 13.

49. B.H. Kolner and D.M. Bloom, *IEEE J. Quantum Electron.* **QE-22**, 79 (1986).

50. E.E. Mendez, L. Esaki, and W.I. Wang, *Phys. Rev. B* **33**, 2893 (1986).

51. V.J. Goldman, D.C. Tsui, and J.E. Cunningham, *Phys. Rev. B* **36**, 7635 (1987).

52. C.E.T. Gonçalves da Silva and E.E. Mendez, *Phys. Rev. B* **38**, 3994 (1988).

53. M.L. Leadbeater, L. Eaves, P.E. Simmonds, G.A. Toombs, F.W. Sheard, P.A. Claxton, G. Hill, and M.A. Pate, *Solid State Electron.* **31**, 707 (1988).
54. M.L. Leadbeater, E.S. Alves, L. Eaves, M. Henini, O.H. Hughes, A. Celeste, J.C. Portal, G. Hill und M.A. Pate, *Phys. Rev.* B **39**, 3438 (1989).
55. C.H. Yang, M.J. Yang, and Y.C. Kao, *Phys. Rev.* B **40**, 6272 (1989).
56. G.S. Boebinger, A.F.J. Levi, S. Schmitt-Rink, A. Passner, L.N. Pfeiffer, and K.W. West, *Phys. Rev. Lett.* **65**, 235 (1990).
57. J.G. Chen, C.H. Yang, M.J. Yang, and R.A. Wilson, *Phys. Rev.* B **43**, 4531 (1991).
58. M.L. Leadbeater, F.W. Sheard, and L. Eaves, *Semicond. Sci. Technol.* **6**, 1021 (1991).
59. Y. Galvao Gobato, T. Chevoir, J.M. Berroir, P. Bois, Y. Guldner, J. Nagle, J.P. Vieren, and B. Vinter, *Phys. Rev.* B **43**, 4843 (1991).
60. T.K. Higman, M.E. Favaro, L.M. Miller, M.A. Emanuel, and J.J. Coleman, *Appl. Phys. Lett.* **54**, 1751 (1989).
61. W. Müller, H.T. Grahn, R.J. Haug, and K. Ploog, *Phys. Rev.* B **46**, 9800 (1992).
62. W. Müller, H.T. Grahn, K. von Klitzing, and K. Ploog, *Surf. Sci.* **305**, 380 (1994).
63. U. Ekenberg, *Phys. Rev.* B **40**, 7714 (1989).
64. F. Malcher, G. Lommer, and U. Rössler, *Superlattices Microstruct.* **2**, 267 (1986).
65. S. Das Sarma, *Phys. Rev. Lett.* **52**, 859 (1984).
66. D.M. Larsen, *Phys. Rev.* B **30**, 4807 (1984).
67. F.M. Peeters and J.T. Devreese, *Phys. Rev.* B **31**, 3689 (1985).
68. R.E. Camley and D.L. Mills *Phys. Rev.* B **29**, 1695 (1984).
69. A.K. Sood, J. Menéndez, M. Cardona, and K. Ploog, *Phys. Rev. Lett.* **54**, 2115 (1985).
70. C. Colvard, T.A. Gant, M.V. Klein, R. Merlin, R. Fisher, H. Morkoç, and A.C. Gossard, *Phys. Rev.* B **31**, 2080 (1985).
71. C. Colvard, R. Merlin, M.V. Klein, and A.C. Gossard, *Phys. Rev. Lett.* **45**, 298 (1980).
72. J.C. Maan, in *Festkörperprobleme 27*, edited by P. Grosse (Vieweg, Braunschweig, 1987), p. 137.
73. W. Müller, H.T. Grahn, K. von Klitzing, and K. Ploog, *Phys. Rev.* B **48**, 11176 (1993).
74. H. Schneider, H.T. Grahn, K. von Klitzing, and K. Ploog, *Phys. Rev.* B **40**, 10040 (1989).
75. A.J. Shields, M. Cardona, H.T. Grahn, and K. Ploog, *Phys. Rev.* B **47**, 13922 (1993).
76. E.E. Mendez, W.I. Wang, B. Ricco, and L. Esaki, *Appl. Phys. Lett.* **47**, 415 (1985).
77. K. Leo, J. Shah, J.P. Gordon, T.C. Damen, D.A.B. Miller, C.W. Tu, and J.E. Cunningham, *Phys. Rev.* B **42**, 7065 (1990).

78. M. Nido, M.G.W. Alexander, W.W. Rühle, and K. Köhler, *Phys. Rev. B* **43**, 1839 (1991).

79. B.T. Norris, N. Vodjdani, B. Vinter, E. Costard, and E. Böckenhoff, *Phys. Rev. B* **43**, 1867 (1991).

80. A. Cros, A. Cantarero, C. Trallero-Giner, and M. Cardona, *Phys. Rev. B* **45**, 6106 (1992).

CHAPTER 5

ELECTRIC FIELD DOMAINS

by HOLGER GRAHN

5.1. Introduction

In the previous chapter it was shown that semiconductor superlattices can exhibit several regions of negative differential velocity, when an electric field is applied parallel to the superlattice direction. The presence of one regime of negative differential velocity (NDV), e.g. in bulk GaAs through the transfer from lighter electrons in the Γ valley to heavier electrons in the L valley,[1] leads to the formation of propagating electric field domains.[2,3] Since in bulk GaAs at high electric fields the drift velocity decreases with increasing field, an instability of the field distribution near one of the contacts will grow in time. The accumulated space charge, which appears due to the local change in the electric field, will move through the structure with a velocity close to the saturation velocity of the corresponding carriers forming a propagating high-field domain.

NDV can result in negative differential resistance (NDR) or conductivity (NDC) of the I-V characteristic. NDR or NDC can also be caused by mechanisms other than NDV, e.g. the field dependence of the carrier density. In CdS single crystals the formation of stable domains on a macroscopic length scale has been observed due to the presence of NDC.[4]

In systems exhibiting NDV the field dependence of the drift velocity is the essential ingredient for domain formation. The continuity and Poisson's equations have to be solved self-consistently in order to obtain the field distribution. In systems with no more than one NDV region only nonstationary solutions exist. However, in systems, where the NDC is caused

by the field dependence of the carrier density, the drift velocity is assumed
to be linearly field dependent, i.e., the transport is described by a field
independent drift mobility. In this case the coupled system of differential
equations leads to a solution with stable electric field domains, i.e., two
regions with constant, but different field strength are observed separated by
a space charge layer, which forms the domain boundary. The corresponding
I-V characteristic shows first a linear increase of the current followed by a
wide plateau, in which the domains are present.

Electric field domains in semiconductor superlattices originate from
the presence of NDV in the drift velocity-field characteristic, but appear
for large carrier densities in the stable form. In very strongly coupled
superlattices with a wide miniband similar domains as in the bulk case are
expected. But in a weakly coupled superlattice with more than one subband
below the barrier as well as a large carrier density a different type of domain
formation appears. Since in a weakly coupled superlattice the carriers can
only be located in the wells, the system is spatially inhomogeneous being
substantially different from bulk GaAs or CdS. Furthermore, electric field
domains are observed on a microscopic length scale ($\sim$ 100 nm) in this
system compared to the macroscopic length scales in bulk semiconductors.

The first observation of electric field domains in superlattices was
reported by Esaki and Chang in 1974.[5] The conductivity of a doped GaAs-
AlAs superlattice exhibited oscillations as a function of the applied voltage,
which were correlated with the spacing of the first two electronic subbands
of this system. The origin and the actual field distribution was discussed in a rather speculative fashion. It took another 10 years, before electric field domains were investigated in different material systems such as
$In_xGa_{1-x}As$-$In_xAl_{1-x}As$,[6,7] GaAs-$Al_{1-x}Ga_xAs$,[8-25] and $In_xGa_{1-x}As$-InP
superlattices[26-28] using transport and optical experiments. The I-V characteristic or conductivity usually exhibits periodic structures as a function
of the applied voltage. The period is connected to spacing between the first
two electronic subbands indicating a relation to resonant tunneling between
the first and second subband. The first experiments on domain formation
with optical excitation demonstrated the existence of well-defined field regions in this system.[14,16]

5.2. Theoretical model for domain formation

Electric field domains appear in semiconductor superlattices due to the non-linear drift velocity-field characteristic (as discussed in Chapter 4) in conjunction with a large carrier density. In weakly coupled superlattices the drift velocity exhibits maxima at field strengths, which correspond to resonant coupling between different subbands in adjacent wells, i.e., $F_i = (E_i - E_1)/(e\ d)$, where E_i denotes the energy of the ith subband and d the superlattice period. In principle these resonances can occur between conduction band or valence band states, but we will limit the discussion to conduction band states. Under steady state conditions the two equations, which govern the transport and field distribution, are current conservation

$$j(F) = e\ n(F)\ v_d(F) = const. ,\qquad(5.1)$$

where j denotes the current density, $n(F)$ the carrier density, and $v_d(F)$ the drift velocity, and Poisson's equation

$$\frac{dF}{dz} = \frac{e\ n(F)}{\epsilon_0\ \epsilon} ,\qquad(5.2)$$

where z is the spatial coordinate parallel to the superlattice direction and ϵ_0 and ϵ are the dielectric constant of the vacuum and the material, respectively. These two equations are coupled and non-linear due to the non-linear dependence of the drift velocity on the electric field. Eqs. (5.1) and (5.2) have to be solved self-consistently. This can be done by replacing the carrier density from Eq. (5.2) in Eq. (5.1) resulting in the combined equation

$$\frac{dF}{dz}\ v_d(F) = const. = j_0 = \frac{j}{\epsilon_0\epsilon} .\qquad(5.3)$$

This equation together with the boundary conditions $V(z = 0) = 0$ and $V(z = L) = F_{ap}L$ determine the field distribution. $V(z)$ is the potential distribution of the conduction band edge, F_{ap} the electric field strength which corresponds to the applied voltage divided by the length of the system in the case of a homogeneous voltage distribution (constant field strength), and L the total thickness of the superlattice, i.e., $L = Nd$, where N corresponds to the number of periods of the superlattice. If no charge is present in the system, the field strength is independent of z according to Eq. (5.2) and the two equations are independent. If the drift velocity is constant, however, the electric field will vary linearly with z resulting in a quadratic dependence of the voltage distributon. This is typical for a charged system. If the drift velocity exhibits resonances, it is possible to get several

solutions, which are stable. We have neglected the time dependent part
of the continuity equation in order to focus on stable solution of the field
distribution. The simplest relation to assume between drift velocity and
electric field in the case of resonant tunneling is a series of δ-functions, i.e.,

$$v_d(F) = \sum_{i=1}^{m} v_i \delta(F - F_i) \,, \qquad (5.4)$$

where v_i is the maximum drift velocity of the ith resonance and m is the
number of subbands below the barrier. If we confine the discussion to two
subbands, Eq. (5.3) is solved in conjunction with Eq. (5.4) resulting in the
function $z(F)$. After inverting this relation into $F(z)$ and imposing the
boundary condition, the result is

$$F(z) = F_1 \, \Theta(z) \, \Theta(z_1 - z) + F_2 \, \Theta(z - z_1) \, \Theta(L - z) \qquad (5.5)$$

with $\Theta(z)$ denoting the unit step function and $z_1 = L(1 - F_{ap}/F_2)$. The applied electric field breaks up into two regions with well-defined field strengths
F_1 and F_2 separated by a boundary at z_1, which is connected with an accumulation of charge. Increasing the applied field only changes the location
of the domain boundary, but not the field strengths which are present in
the superlattice. This is very different from the extremely low carrier density case, where the field is homogeneously distributed and increases with
increasing bias. Due to the discreteness of the system, the space charge
at the domain boundary can only be located within a well. Therefore, the
domain boundary moves discontinuously through the superlattice. In this
simple model the domain boundary is reduced to one point in space. In
reality it will distributed over at least one period.

If more than two subbands are taken into account, a possible solution
of Eq. (5.3) can contain three coexisting domains exhibiting field strengths
F_1, F_2, and F_3, if

$$\frac{v_2}{j_0} \leq \frac{F_3 - F_{ap}}{F_3 - F_2} L = z_2 \,, \qquad (5.6)$$

resulting in the following field distribution

$$\begin{aligned} F(z) = F_1 \, \Theta(z) \, \Theta(z_3 - z) \quad &+ F_2 \, \Theta(z - z_3) \, \Theta(z_4 - z) \\ &+ F_3 \, \Theta(z - z_4) \, \Theta(L - z) \end{aligned} \qquad (5.7)$$

with

$$z_3 = (1 - \frac{F_{ap}}{F_3})L - \frac{v_2}{j_0}(1 - \frac{F_2}{F_3}) \qquad (5.8)$$

and

$$z_4 = (1 - \frac{F_{ap}}{F_3})L + \frac{v_2}{j_0}\frac{F_2}{F_3} \tag{5.9}$$

Now there are two domain boundaries at z_2 and z_3, which are spatially ordered according to an increasing field strength. The two field distributions of Eqs. (5.5) and (5.7) are schematically shown in Fig. 5.1.

In this oversimplified model the domain boundary is infinitesimally small. In the drift velocity-field characteristic of an experimentally realized superlattices the δ-function like resonances are broadened. At the same time the peak value of the resonance increases with increasing index i. A schematic characteristic is shown in Fig. 5.2. Due to current conservation the low-field domain will experience a resonance field strength, while the high-field domain will be below resonance as shown in the inset. The same picture applies for the case, when resonant coupling between $C1 \rightarrow C2$ and

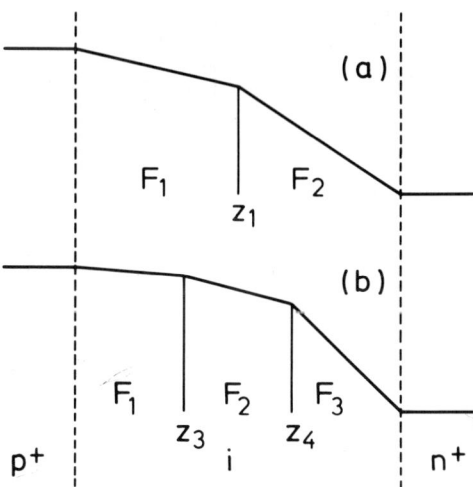

Fig. 5.1. Potential distribution of two (a) and three (b) coexisting domains. The domain boundary is indicated by vertical lines. The field strengths F_1 and F_2 in (a) and (b) are different.

$C1 \rightarrow C3$ is considered. The field strength of the low-field domain experiencing F_I is not zero, since the current would be zero in this case. The current increases linearly at very low fields (cf. Chapter 2), exhibits a maximum at intermediate fields, and subsequently decreases again, before the increase to the resonance between $C1$ and $C2$ begins. The first maximum occurs in superlattices with a substantial miniband width due to miniband transport as described in Chapter 2. For superlattices with a miniband width below the level broadening it appears due to the detuning of the levels between adjacent wells out of resonance.

The first theoretical investigation on the possibility of an electrical instability in a superlattice was reported by Suris.[29] In a subsequent paper propagating Gunn domains were evaluated in superlattices.[30] Instabilities of an applied electric field were also studied theoretically by Ignatov et al.[31] However, no clear indication of large stable domains was found.

There have been a number of theoretical investigations of electric field

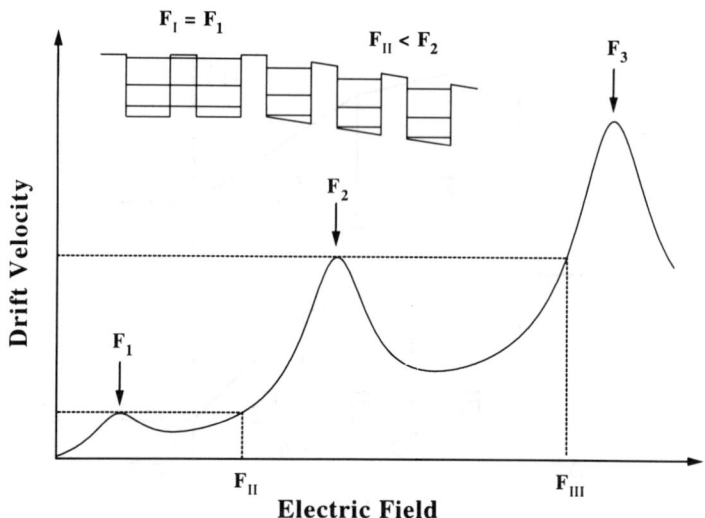

Fig. 5.2. Schematic drift velocity-field characteristic of an undoped superlattice with a negligible carrier density. The three peaks at F_1, F_2, and F_3 are due to the resonant alignment of $C1 \rightarrow C1$, $C1 \rightarrow C2$, and $C1 \rightarrow C3$, respectively. Inset: Schematic diagram showing the coexistence of domains I and II.

domain formation in semiconductor superlattices beyond the simple model described above. Most studies used the drift velocity-field characteristic as shown in Fig. 5.2 as a starting point. Laikhtman[32] applied a circuit theory approach to calculate the voltage distribution in a superlattice under domain formation. In a subsequent publication[33] the stability of the miniband regime of a superlattice was investigated using a coherent tunneling picture. However, no higher mini- or subbands were included. An analytical, microscopic model, which includes two narrow minibands and assumes a complete relaxation between the second and first miniband, was recently employed by the same authors[34] to determine the field and carrier distribution as well as the current-voltage characteristic. It was found that the first peak in the domain regime exhibits the largest current and that the voltage spacing of the branches is below the subband spacing. Recently, a microscopic model of domain formation, which assumes miniband transport in the lowest subband and resonant tunneling between the first and second subband, was developed by Prengel et al.[35−38] for a doped superlattice. This model is based on a numerical, self-consistent solution of the rate equations and Poisson's equation. The field and carrier distribution as well as the static I-V characteristic were calculated in the low and high carrier density limit. Bonilla et al.[39−41] studied the transition from unstable to stable field distributions as a function of the carrier density in a photoexcited system, i.e., taking into account electrons and holes, using a phenomenological model. If the carrier density is not large enough to form stable domains, the domain boundary will oscillate over several periods. Single electron quantization of electric field domains in slim semiconductor superlattices was theoretically investigated by Korotkov et al.[42] using a Monte Carlo simulation.

In Fig. 5.3 the I-V characteristic of a GaAs-AlAs superlattice with 20 periods, 9.0 nm GaAs wells, 1.5 nm AlAs barriers, and a two-dimensional carrier density of 4.8×10^{10} cm^{-2} calculated using the model by Prengel et al.[35] is shown for two sweep directions. In the upper part the solid lines correspond to a sweep from 0 to 3.0 V, while the lower solid lines show the sweep from 3.0 to 0 V. The I-V consists of a plateau, i.e., the average current is almost constant, with 20 jumps, which are infinitesimally sharp. This model predicts that the domain boundary is more or less confined to one well. A closer look at the field distribution shows that the domain boundary extends over three periods, but 90% of the field change occurs

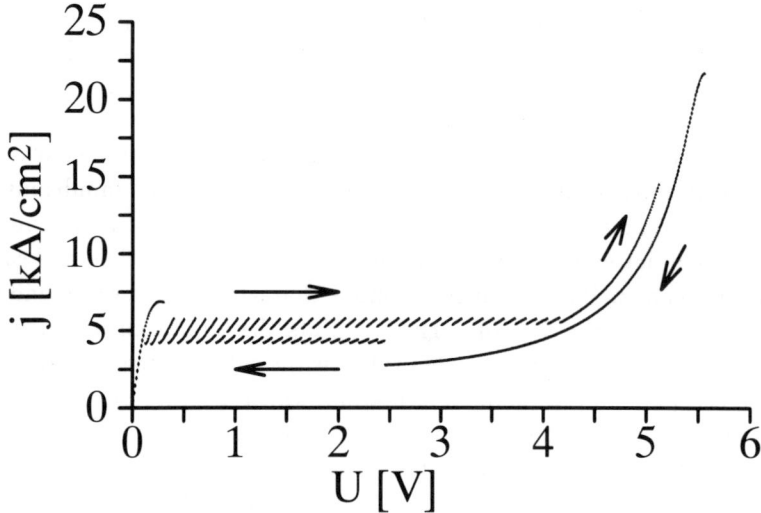

Fig. 5.3. Calculated I-V characteristic for a GaAs-AlAs superlattice with 9.0 nm wells, 1.5 nm barriers, 20 periods, and a carrier density of 4.8×10^{10} cm^{-2}. The upper and lower solid lines correspond to a sweep-up and a sweep-down, respectively, of the bias voltage. The dashed lines correspond to intermediate stable branches (from Ref. 38).

over one period. Due to the discreteness of the superlattice this type of behavior is expected even within the simple model described above. The calculated I-V characteristics exhibits hysteresis, which leads to a multistability of the current for a fixed applied voltage, which has been observed in the I-V characteristic of doped superlattice structures.[25]

5.3. Current-voltage characteristics

Electric field domain formation in superlattices is linked to the presence of a large carrier density. The carriers can be introduced by doping the superlattice, which leads to the transport of either electrons or holes. Another method to introduce the carriers into the superlattice is to photoexcite carriers in an undoped superlattice leading to the simultaneous presence of electron and holes within the superlattice. Doping has the advantage of having only one type of carrier (electron or hole) present in the system. The advantage of photoexcitation originates from the possibility of changing the carrier density over several orders of magnitude within the same sample in order to study the dependence on the carrier density. However, the simultaneous presence of electron and holes leads to a more complicated I-V characteristic.

In this section we will discuss first the domain formaton in doped superlattices, followed by a discussion on the multistability of the I-V characteristic. We will then present the domain formation in undoped superlattices under photoexcitation and compare the resulting I-V characteristic with the characteristic of the doped superlattices.

5.3.1. *Doped superlattices*

The doped GaAs-AlAs superlattices are grown in a n^+-n-n^+ configuration with the Si-doping being confined to the central 5 nm of the GaAs wells. The barriers are nominally undoped. The superlattice forms the n-layer. Typical I-V characteristics of an undoped and a doped superlattice with the same well and barrier thickness are shown in Fig. 5.4. The undoped superlattice (Fig. 5.4(a)) does not exhibit domain formation in this voltage range, while the I-V characteristic of the doped superlattice (Fig. 5.4(b)) clearly displays domain formation. The current increases strongly at low voltages followed by a plateau-like region between -0.5 and -4.5 V, which exhibits a lot of fine structure. There are 36 discontinuites in the I-V characteristic for a superlattice with 40 periods. Each discontinuity corresponds to the movement of the domain boundary by one period. Since the number of jumps is less than the number of periods, the domain boundary might move more than one period, when it is located near one of the contacts. The spacing of these jumps is on the average 110 meV, which is somewhat smaller than the subband spacing of 135 meV. This observation is in agreement with the picture described later, where it is shown that the field strength of the high-field domain is actually below resonance. The discontinuities are infinitesimally sharp, since even with a voltage resolution of 1 μV no finite width is resolved. The movement of the domain boundary also leads to giant capacitance oscillations as has been reported recently by Zhang et al.[43] The presence of two well-defined field regions will be demonstrated in the section on photoluminescence spectroscopy.

The spatial extent of the domain boundary is of the order of 1 period.

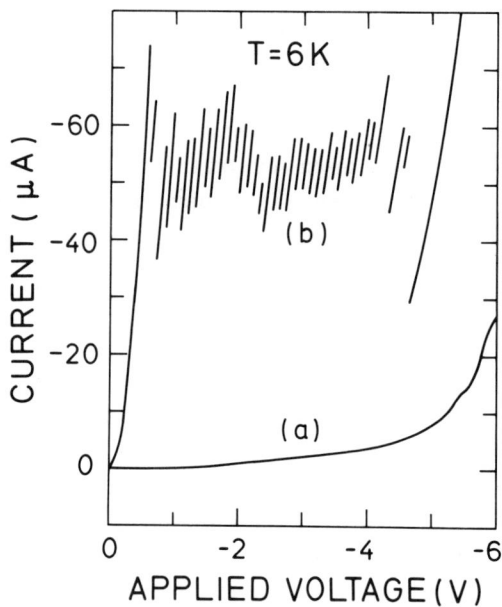

Fig. 5.4. Current-voltage characteristic of a nominally undoped (a) and a doped (b) GaAs-AlAs superlattices with 9.0 nm GaAs layers, 4.0 nm AlAs layers, and a Si-doping density of 3×10^{17} cm^{-3} at 5 K.

This can be extracted by analyzing at the conductance of each of the branches in the I-V characteristic. At the domain boundary the coupling will be non-resonant. We can therefore use the model for non-resonant tunneling, which contains an exponential dependence, to estimate the conductance. Because the current depends through an exponential function on the applied voltage, we will use the derivative of the logarithm of the current, which is equal to the differential conductance normalized by the current. In Fig. 5.5 the numerical derivative of the experimental data of Fig. 5.4(b) is plotted versus the applied voltage. Each discontinuity appears as a sharp negative spike. What is important in this figure, is the value of $dln(I)/dV$ between the spikes, which corresponds to the slope of the branches. The slope varies between 2.9 V^{-1} at -0.6 V and 1.3 V^{-1} at -4.6 V. If we use the semi-classical model for non-resonant tunneling[44] at the domain boundary,

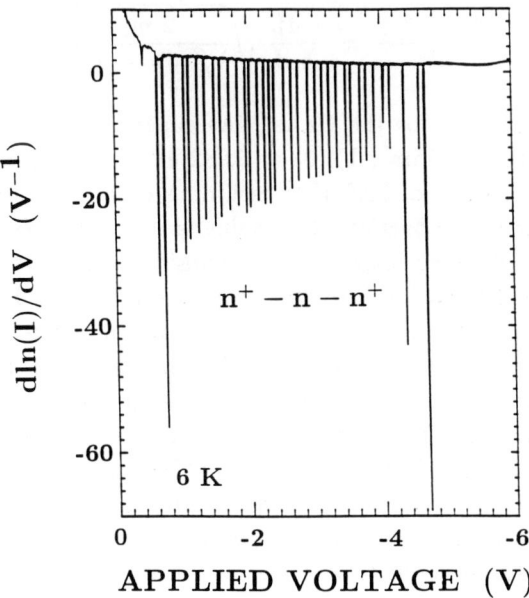

Fig. 5.5. Numerical Derivative of the I-V characteristic of the doped GaAs-AlAs superlattices with 9.0 nm GaAs layers, 4.0 nm AlAs layers, 40 periods, and a doping density
of 3×10^{17} cm^{-3} at 5 K vs applied voltage. The corresponding I-V characteristic is
shown in Fig. 5.4.

the current should be described by

$$I = I_0 \, exp(-2 \, d_B \, \sqrt{2 \, m_B^* \, (V_B^C - E_{C1})}/\hbar) \,, \qquad (5.10)$$

where d_B denotes the barrier thickness, m_B^* the electron effective mass in
the barrier, V_B^C the barrier height, and E_{C1} the binding energy of the lowest
conduction subband. Increasing the electric field on the superlattice leads
only to a decrease of the effective barrier height at the domain boundary,
i.e., $V_B^C(F) = V_B^C(0) - eF d_B$. The applied voltage at the domain boundary
is equal to the total applied voltage minus the number of periods in the
high-field domain times the resonance field strength of the second domain
times the superlattice period. Putting this together, the derivative can be

approximated by

$$\frac{dln(I/I_0)}{dV} = \frac{e\,d_B^2}{\hbar\,d}\sqrt{\frac{2\,m_B}{V_B^C(0)\,-\,E_{C1}}}. \tag{5.11}$$

Using the parameters for the GaAs-AlAs superlattice, i.e., $d_B = 4$ nm, $d = 13$ nm, $m_B^* = 0.15m_0$, $V_B(0) = 0.982$ eV, and $E_{C1} = 44$ meV, the derivative has a value of 2.52 V^{-1} assuming that the domain boundary extends only over one barrier. This value is in good agreement with the experimentally observed slope. This conclusion is supported by the field distribution, which can be extracted from the model by Prengel et al.[35] The corresponding I-V characteristic was shown in Fig. 5.3 in qualitative agreement with the experimental one. The modulation of the plateau region in the experiment is expected to arise from inhomogeneities in the superlattice growth. This assignment is under current investigation applying the microscopic model by Prengel et al.[35] The field distribution which is obtained from the calculations shows that the domain boundary is confined to about 90% within one superlattice period, i.e., most of the accumulated charge at the domain boundary resides in one well.

A comparison of the I-V characteristics from three different samples is shown in Fig. 5.6. The negative voltage region in (a) has already been discussed in connection with Fig. 5.4. However, it seems that no domain formation is found in the positive voltage region. It turns out that the current is not stable in this regime, but oscillates at a frequency of a few hundred kHz. These oscillatons will be discussed in the last section of this chapter. The trace in (b) was obtained on a sample with the same doping as in (a), but with a different top contact layer. For this sample the current shows domain formation in both bias directions. For a doping of 1×10^{18} cm^{-3} with the same contacts as for the sample in Fig. 5.6(a) the domain formation also clearly exists in both bias direction (Fig. 5.6(c)). We conclude that the carrier density should exceed a certain value in order to observe stable domain formation in the I-V characteristic. For the I-V characteristic in Fig. 5.6(a) the top contact does not supply enough carriers in order to achieve stable domains, while in the other two samples either the contact supplies enough carriers (Fig. 5.6(b)) or the doping is sufficiently large (Fig. 5.6(c)). We will return to the positive voltage region of the I-V characteristic in Fig. 5.6(a) in the section on photoluminescence spectroscopy and the dynamics of domain formation.

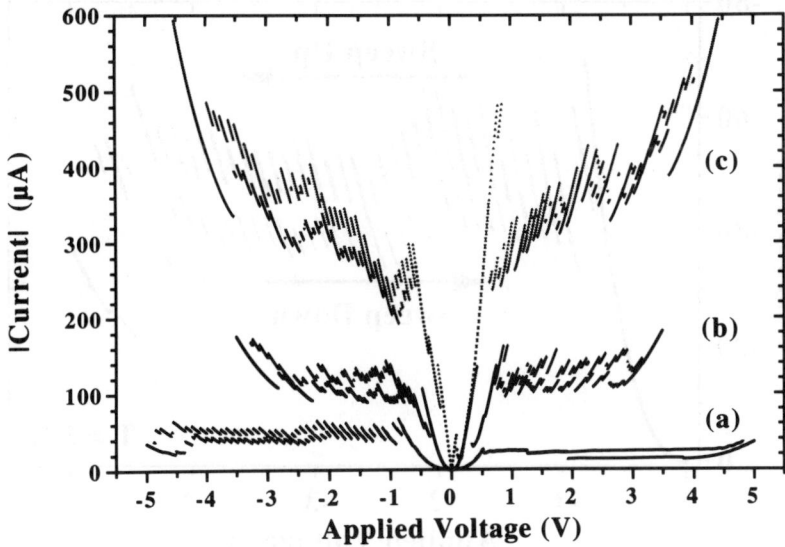

Fig. 5.6. Current-voltage characteristic of three doped GaAs-AlAs superlattices with 9.0 nm GaAs layers, 4.0 nm AlAs layers, and 40 periods for both bias directions at 5 K. The doping density is 3×10^{17} cm^{-3} in (a) and (b), while it is 1×10^{18} cm^{-3} in (c). The difference between (a) and (b) is due to a different top contact layer.

5.3.2. *Multistability*

The calculated I-V characteristic in Fig. 5.3 displays a strong hysteresis. This asymmetry between a sweep-up and a sweep-down is also observed in the experiment. In Fig. 5.7 a sweep-up from 0 to −5 V and a sweep-down from −5 to 0 V is shown for the GaAs-AlAs superlattice with a doping density of 3×10^{17} cm^{-3}. As in the calculated curve the sweep-up leads to a higher current level than the sweep-down. However, the difference between sweep-up and sweep-down is more pronounced in the calculated I-V characteristic. In both sweep directions there are 36 discontinuities in the experimental curve. Because each discontinuity reflects the charging or discharging of the well at the domain boundary, a hysteresis effect is expected. A closer look at the data reveal that the upper and lower branches are connected, i.e., each branch of the sweep-down has a counterpart in the

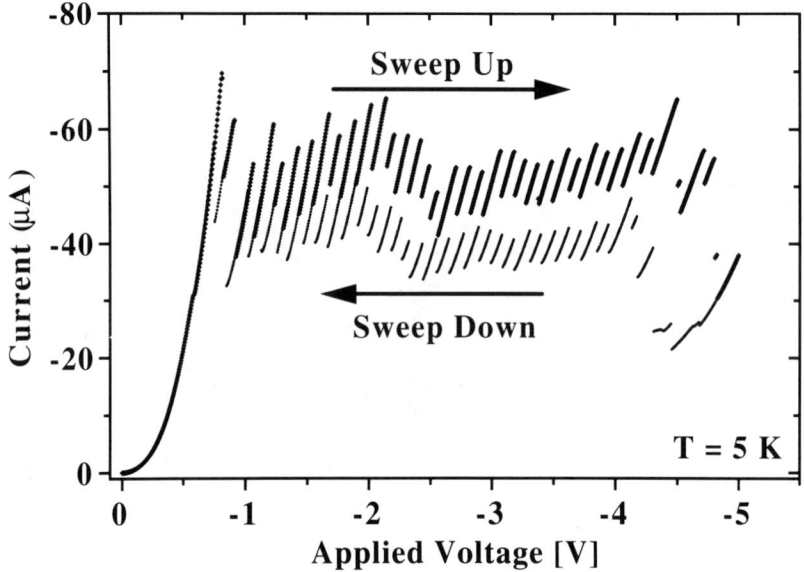

Fig. 5.7. Current-voltage characteristic of the doped GaAs-AlAs superlattices with 9.0 nm GaAs layers, 4.0 nm AlAs layers, 40 periods, and a Si-doping density of 3×10^{17} cm^{-3} at 5 K for two sweep directions. The sweep-up was taken from 0 to -5 V, while the sweep-down went from -5 to 0 V (from Ref. 25).

sweep-up. While in the experiment between 0 and -2 V the two sweep directions clearly overlap, a gap opens up between the two sweep directions for voltages between -2 and -4.8 V. From the hysteresis of the I-V characteristic it is clear that the current can exhibit multistability within the domain regime. In Fig. 5.8(a) an enlarged section of the I-V characteristic of Fig. 5.7 in the region with overlapping branches is shown. The sweep direction is indicated by arrows. The turning point of the sweep direction is in this case -1.6 V. There are regions, where only one branch is present at a given bias voltage (e.g., between -1.25 and -1.33 V) corresponding to monostable operating points. Between -1.33 and -1.42 V, however, the two current branches are clearly separated and the sample exhibits bistability. This bistability occurs, because for each current branch the domain boundary is located in a different well. Thus having two separated current branches at a given bias voltage means that depending on the sweep history

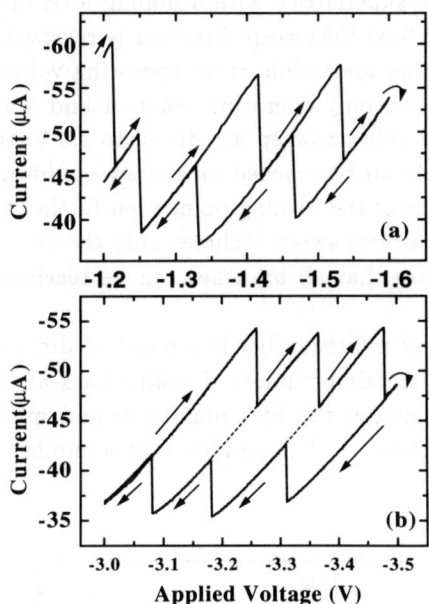

Fig. 5.8. Two enlarged sections of the I-V characteristic at 5 K as shown in Fig. 5.7. In (a) a bistable region is shown, while in (b) the tristable regime is presented. The dashed lines in (b) have been added to indicate the connection of the current branches (from Ref. 25).

the domain boundary can be found in one of two adjacent wells. The I-V characteristic of Fig. 5.7 is shown in Fig. 5.8(b) on an enlarged scale in the regime, where the two branches from the sweep-up and sweep-down are disconnected. The turning point in this case is −3.5 V. Sweep-up and sweep-down do not conincide between −3.1 and −3.45 V, but as indicated by the dashed lines between the upper and lower branches, there is a close connection between these branches. The two branches, which are connected by the dashed lines, belong actually to a single branch, which cannot be mapped out completely in this type of sweep. The dashed region at −3.2 V can be reached by reversing the sweep direction at, e.g., −3.15 V. In this case the current would evolve exactly along the dashed line towards the upper part of the sweep.

This is demonstrated in Fig. 5.9, where an enlarged section of the I-V characteristic of the superlattice with a doping level of $(1 \times 10^{18} \text{ cm}^{-3})$ is shown. In Fig. 5.9(a) the sweep direction is reversed at 3.7 V. A total of four stable branches are visible at an operating voltage of 3.5 V. In the displayed sweep mode, only operating points 1 and 4 are reached. If the turning point of the voltage sweep is reduced to 3.6 V, the operating point 3 instead of point 4 can be reached on the sweep-down (cf. Fig. 5.9(b)). Reducing the voltage of the turning point even further to 3.55 V as shown in Fig.5.9(c) the complete sweep includes only the operating points 1 and 2. This demonstrates that all branches can be reached by a well-defined voltage sweep.

The presence of multistability is a result of different possibilities for the location of the domain boundary. Looking back at Fig. 5.2 the allowed current can vary between the first maximum and minimum in the drift velocity-field characteristic. This implies that a number of field strengths

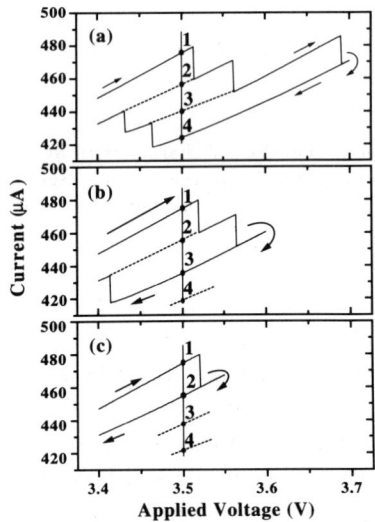

Fig. 5.9. (a) I-V characteristic of the GaAs-AlAs superlattice with a Si-doping density of $1 \times 10^{18} \text{ cm}^{-3}$ at 5 K showing four stable operating points. Each of the indicated points (1, 2, 3, 4) can be reached by a well-defined voltage sweep as shown in (b) and (c) (from Ref. 25).

for the low and high field domain are possible as long as current conservation is fulfilled. In the simplest form the applied voltage is divided up between the low- and high-field domain

$$V_{ap} = n F_1 d + (N - n) F_2 d . \qquad (5.12)$$

where n denotes the number of periods in the low-field domain. There are several solutions to Eq. (5.12) possible, if we allow a small change of the electric field strengths of the low- and high-field domain. If the low-field domain increases by one period keeping the applied voltage fixed, the field strengths F_1 and F_2 are changed to F_1' and F_2'. These new field strengths have to compatible with current conservation and a constant voltage. The new field strength F_2' can be estimated using Eq. (5.12) together with current conservation and the boundary condition

$$F_2' = \frac{N - n}{N - n - 1} F_2 + \frac{n}{N - n - 1} F_1 - \frac{n + 1}{N - n - 1} F_1' . \qquad (5.13)$$

The last two terms are negligible compared to the first one. The field

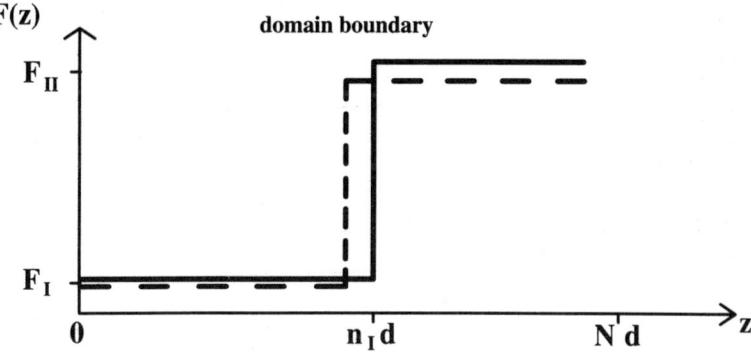

Fig. 5.10. Schematic field distribution within the superlattice for different current branches at a fixed applied voltage.

strength of the high-field domain is therefore slightly increased in order to achieve the same applied voltage. The field strength of the low-field domain also has to be increased slightly due to current conservation. This is schematically shown in Fig. 5.10 for three different possibilities for the location of the domain boundary. This picture is supported by calculations of the field distribution within the microscopic model.[25]

5.3.3. *Undoped superlattices*

In undoped superlattices carriers can only be injected from the contacts or produced by photoexcitation. In order to distinguish between these two effects usually p$^+$-i-n$^+$ structures are biased in the reverse direction so that no carriers are injected via the contacts. A laser beam with a photon energy above the bandgap of the superlattice is focused onto the sample, and electrons as well as holes are excited. In Fig. 5.11 the photocurrent-voltage characteristic in reverse bias of an undoped GaAs-AlAs superlattice

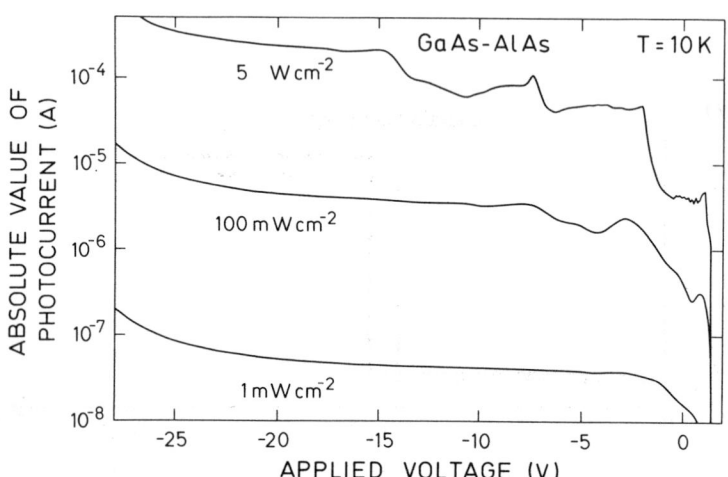

Fig. 5.11. Absolute value of the photocurrent vs applied voltage of a GaAs-AlAs superlattice with 14.4 nm GaAs wells, 3.4 nm AlAs barriers and 50 periods at 10 K for different excitation densities at 1.959 eV (632.8 nm).

with 14.4 nm GaAs wells, 3.4 nm AlAs barriers, and 50 periods is shown for different laser intensities, i.e., carrier densities. While the curves at 1 and 100 mW cm^{-2} show very little structure except for some tunneling resonances as discussed in Chapter 4, the trace at 5 W cm^{-2} shows already plateau-like regions with some fine structure. Between the plateaus the current increases very strongly. However, no clear evidence of domain formation is seen yet. Increasing the laser intensity by another factor of ten leads to the photocurrent-voltage characteristic shown in Fig. 5.12. The sample now clearly exhibits domain formation within the regions A, B, C, and D, which are all plateau-like containing a lot of fine structure. The number of discontinuities is in the range of 20 to 25, which is significantly less than the number of periods. The forward bias direction is also included in Fig. 5.12 to demonstrate the importance of photoexcitation even in the injection regime. The regions B and C correspond to the coexistence of F_1 and F_2, while the regions A and D now denote coexisting domains of F_2

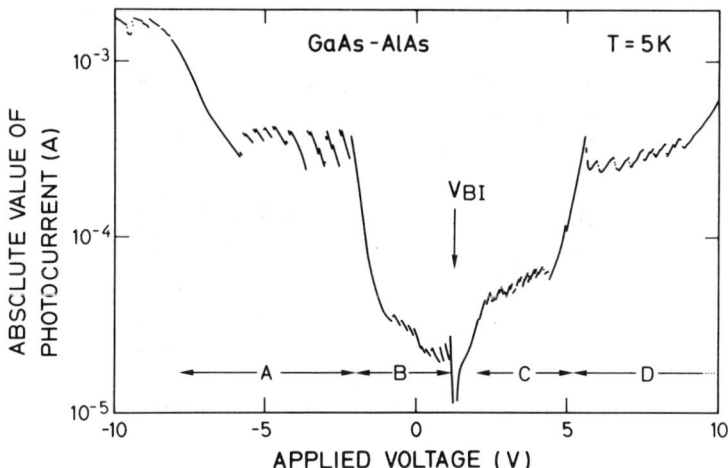

Fig. 5.12. Absolute value of the photocurrent vs applied voltage of the GaAs-AlAs superlattice with 14.4 nm GaAs wells, 3.4 nm AlAs barriers, and 50 periods at 10 K using a laser intensity of 50 W cm^{-2} at 1.916 eV (647 nm). The arrow denotes the built-in voltage V_{BI}. The regions labelled A, B, C, and D are discussed in the text.

and F_3. In this sample the subband spacing is smaller than in the previously discussed ones. Consequently, it is easier to observe the regime of even higher domain field strengths. There is even another plateau with fine structure beyond -9 V corresponding to coexisting domains of F_3 and F_4. In forward bias without photoexcitation this sample does not exhibit fine structure in the I-V characteristic a voltage of 8 V.

The effect of the simultaneous presence of electrons and holes can be studied, when a doped structure is photoexcited and its I-V characteristic is compared to the one of the undoped structure. In Fig. 5.13 the photocurrent-voltage characteristic of the undoped superlattice with 9.0 nm GaAs wells, 4.0 nm barriers, and 40 periods in a p^+-i-n^+ structure is shown for a large laser intensity. The built-in voltage ($V_{BI} = 1.5$ V) has been subtracted in order to compare this trace directly with the I-V characteristic of the doped superlattice (n^+-n-n^+ structure) with the same parameters

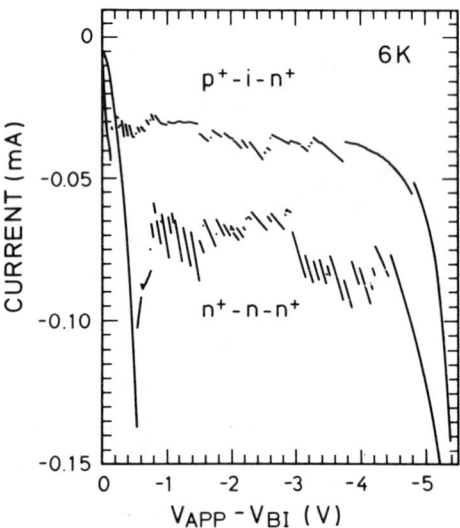

Fig. 5.13. Total current vs applied voltage of GaAs-AlAs superlattices with 9.0 nm GaAs wells, 4.0 nm AlAs barriers, and 40 periods at 6 K using a large laser intensity at 1.916 eV (647 nm). The p^+-i-n^+ structure is undoped, while the n^+-n-n^+ structure contains Si-doping with a density of 3×10^{17} cm^{-3}. The built-in voltage V_{BI} was 1.5 V for the p^+-i-n^+ structure and and 0 V for n^+-n-n^+ structure.

(Fig. 5.13). The fine structure in the I-V characteristic of the p^+-i-n^+ structure is less pronounced and regular than in the I-V trace of the n^+-n-n^+ structure. Furthermore, the I-V characteristics of the doped superlattice loses some of its regularity under strong photoexcitation. Both observations are a direct consequence of the simultaneous presence of the electrons and holes under photoexcitation. The n^+-i-n^+ structure with the same superlattice parameters exhibits a similar I-V characteristic under photoexcitation as the p^+-i-n^+ structure.

The samples with 9.0 nm GaAs wells and 4.0 nm AlAs barriers exhibit a second plateau regime at larger electric fields. In Fig. 5.14 the n^+-i-n^+ structure is shown in this field regime without (a) and with (b) strong illumination. The dark characteristic does not display any signs of domain formation. Under strong illumination, however, a clear signature of domain formation is observed with fairly regular fine structure and a number of discontinuities of the order of the number of periods.

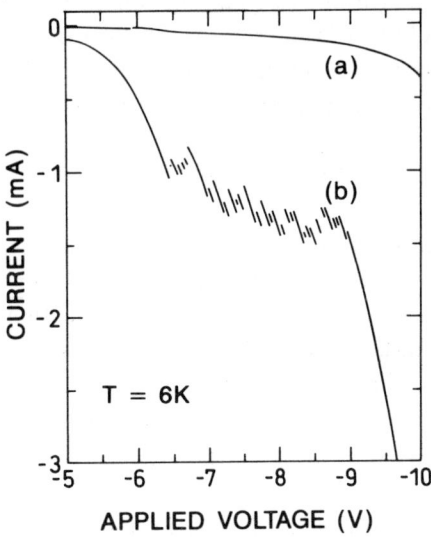

Fig. 5.14. Dark (a) and total current (b) vs applied voltage of the undoped GaAs-AlAs superlattices with 9.0 nm GaAs wells, 4.0 nm AlAs barriers, and 40 periods at 6 K in the n^+-i-n^+ configuration. The excitation conditions in (b) are the same as in Fig. 5.13.

In summary, the doped structures exhibit the clearest evidence of domain formation in the I-V characteristics. Strong photoexcitation of undoped structures (p⁺-i-n⁺ and n⁺-i-n⁺ configuration) also leads to domain formation with less regular fine structure due to the simultaneous presence of electrons and holes. The appearance of regular fine structure correlated with the number of periods is a clear evidence for domain formation. The absence of such fine structure in a heavily doped or photoexcited system cannot be taken as a signature for the absence of domain formation. This conclusion will be supported in the next section, where electric field domains are investigated with optical techniques.

5.4. Photoluminescence spectroscopy of electric-field domains

The transport experiments provide information about the movement of the domain boundary. The identification of two well-defined regions with homogeneous field strengths, i.e., the field domains, can only be achieved by a spectroscopic technique such as photoluminescence or intersubband Raman spectroscopy. The energy levels of quantum well structures are red-shifted in an electric field due to quantum-confined Stark effect (QCSE).[45] The energy shift $\Delta E(F)$ of the ground state is proportional to the square of the applied electric field and the fourth power of the well width d_W as determined by perturbation theory for lower fields

$$\Delta E(F) = -g\, d_W^4\, F^2 \ . \tag{5.14}$$

The constant g is proportional to the effective mass, i.e., the shift is much larger for the heavy-hole valence ground state than for the electronic conduction ground state. Therefore, interband transitions originating from different field strength regions exhibit a larger splitting than intersubband transitions, so that photoluminescence spectroscopy is better suited to study the domain field strengths than intersubband Raman spectroscopy.

In the first part we will prove the existence of two well-defined field regions by detecting the photoluminescence (PL) from the band-gap transition. The field strength of the high-field domain can be determined by looking at the PL signal from a higher subband, which is responsible for the high-field domain. The second part contains the intensity dependence of the PL signal from a photoexcited system. Finally, the spatial ordering of the domains will be briefly discussed.

5.4.1. *Determination of domain field strengths*

The presence of two well-defined regions with constant field strength is directly observed in photoluminescence experiments. In Fig. 5.15 the PL spectra of the doped GaAs-AlAs superlattice with 9.0 nm GaAs wells, 4.0 nm AlAs wells, 40 periods, and a Si-doping of 3×10^{17} cm^{-3} is shown for different applied voltages between -1 and -5 V. The optically excited carrier density is well below the doping level and does not change the I-V characteristic of Fig. 5.4(b). Two distinct PL lines are observed which do not shift in energy as the applied voltage is changed. However, intensity is transferred from the high-energy to the low-energy line as the electric field is increased. The high-energy line decreases in strength as the applied field is increased, while the low-energy line gains intensity. This type of

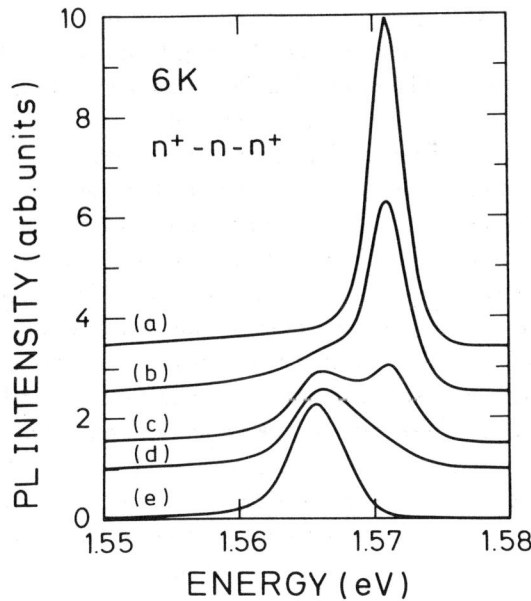

Fig. 5.15. Photoluminescence spectra of the doped GaAs-AlAs superlattices with 9.0 nm GaAs wells, 4.0 nm AlAs barriers, 40 periods, and a Si-doping of 3×10^{17} cm^{-3} at 6 K for different applied voltages. (a) -1 V, (b) -2 V, (c) -3 V, (d) -4 V, and (e) -5 V. The spectra have been shifted vertically for clarity.

spectrum in conjunction with the I-V characteristic clearly demonstrate the existence of two well-defined field domains with a very different field strength. A comparison with the p^+-i-n^+ diode of the same parameters leads to the conclusion that the field strengths of the two domains are in the range of F_1 and F_2. The n^+-i-n^+ structure with the same superlattice parameters shows under weak photoexcitation only a single PL line with a quadratic shift, when the applied field is increased. This is in agreement with the I-V characteristic in Fig. 5.4(a), which does not exhibit any fine structure. In Fig. 5.6(a) the I-V characteristics of the doped superlattice did not display domain formation for positive voltages, although it is expected. The photoluminescence measurements in this voltage regime show clearly two lines similar to the one shown in Fig. 5.15. This results can be understood in terms of an unstable domain boundary which is oscillating. The oscillation of the domain boundary averages out all the fine structure, which is observed for stable domains. However, even if the domain boundary is oscillating over several periods, there can be still large regions of constant field strength covering many periods leading to a similar PL spectrum as for negative voltages. The conclusion from this observation is that neither the I-V characteristic nor the PL experiments alone can give conclusive evidence about the existence of domains. Only both experiments together can give this information. We will see later that the current is actually oscillating in the positive voltage regime supporting this interpretation.

The field strength of the low-field domain is near the maximum of the corresponding resonance according to the simple model described in section 5.2. The high-field domain, however, should experience a field strength below the resonance value, when current conservation and a realistic shape of the resonances is taken into account. The field strength of the high-field domain can be measured by looking at the photoluminescence of the corresponding higher subband. When carriers are injected into a higher subband via resonant tunneling, the recombination from electrons in this higher subband (Cn) with heavy holes in the ground state (H1) can be detected.[46-48] Even in this weakly coupled superlattice the PL signal from the higher subband is observed, although it is about six orders of magnitude smaller than the band-gap PL.[49]

A comparison of the I-V characteristic and the C2H1 PL intensity for the doped GaAs-AlAs superlattice is shown in Fig. 5.16.[24] There is no sign of the C2H1 PL until the current rises to the second plateau. Since the

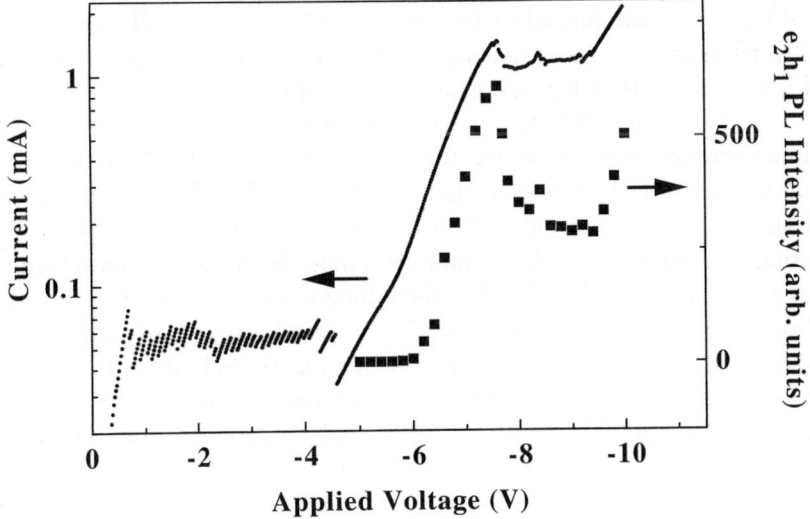

Fig. 5.16. Comparison between C2H1 PL intensity (circles) and total current (small dots) for the doped GaAs-AlAs superlattice with 9.0 nm GaAs wells, 4.0 nm AlAs barriers, 40 periods, and Si-doping of 3×10^{17} cm^{-3} at 5 K (from Ref. 24).

C2H1 PL is detectable only when a resonant alignment of C1 and C2 is achieved, the field strength of the high-field domain in the first plateau region is below the resonance field strength F_2 as expected from current conservation. In the transition regime between the two plateaus the field is homogeneously distributed. At the same time the applied field strength approaches the resonance field strength F_2. This interpretation is in agreement with the qualitative picture shown in Fig. 5.2.

In an undoped GaAs-Al$_{0.35}$Ga$_{0.65}$As superlattice with 13.1 nm wells, 7.9 nm barriers, and 100 periods similar results have been obtained under strong photoexcitation.[24,50] Crossing the second plateau in a voltage sweep, the PL signal from the next higher subband C3, which should determine the field strength of the high-field domain, is not detected until the current strongly increases. In the transition region the C3H1 PL line appears indicating resonant alignment between C1 and C3. The PL from the second subband, C2H1, however, is visible throughout the whole plateau region, decreasing in intensity as the high-field domain grows. Intersubband Raman spectra have been used to determine the subband spacing between C1 and C2 in this sample, when the electric field is applied. The experimentally determined subband spacing agrees very well with the applied voltage, at which the C2H1 PL signal exhibits a maximum.

In the very simple theoretical treatment of domain formation using infinitesimally sharp peaks for the resonances, the field strengths of the domains are given by the resonance field strengths F_i. When the broadening of the resonances and different peak heights are taken into account, the actual field strengths of the domains will differ from the resonance fields as described in section 5.2. The low-field domain experiences a field strength close to the resonance value. However, the field strength of the high-field domain is clearly below the respective resonance field strength, since the current through the sample has to be conserved. We conclude that the low-field domain is resonantly coupled, while in the high-field domain the transport is dominated by non-resonant tunneling.

In the undoped superlattice with the wider well (14.4 nm GaAs) the subband spacings are smaller and electric field domains with field strengths of not only F_1 and F_2, but also F_2 and F_3, and even F_3 and F_4 can be observed. The corresponding I-V characteristic has been shown in Fig. 5.12 displaying three plateau regions with very sharp fine structure indicating domain formation. Only PL experiments in connection with the I-V characteristic can prove the existence of well-defined domains. A question, which was discussed in conjunction with the simple model described in section 5.2, arises with regard to the coexistence of more than two domains for a given voltage. In Fig. 5.17 PL spectra of this sample are shown for applied voltages in region A of Fig. 5.12. Due to the larger fields and well width, the PL lines from the two domains are clearly separated. In this case the high-energy line corresponds to a field strength of F_2, while the low-energy line from the high-field domain is in the range of F_3. Again, the PL lines hardly shift in energy. Only the relative intensity changes.

The PL peak energies of this sample are plotted as a function of the applied voltage in Fig. 5.18. Altogether five different PL lines are observed, which do not shift in energy. These lines can be attributed to the low-field domain F_1 (squares), the high-field and low-field domain F_2 (circles), the high-field and low-field domain F_3 (triangles), and the high-field domain F_4 (diamonds). There is a fifth line (crosses) energetically between the PL

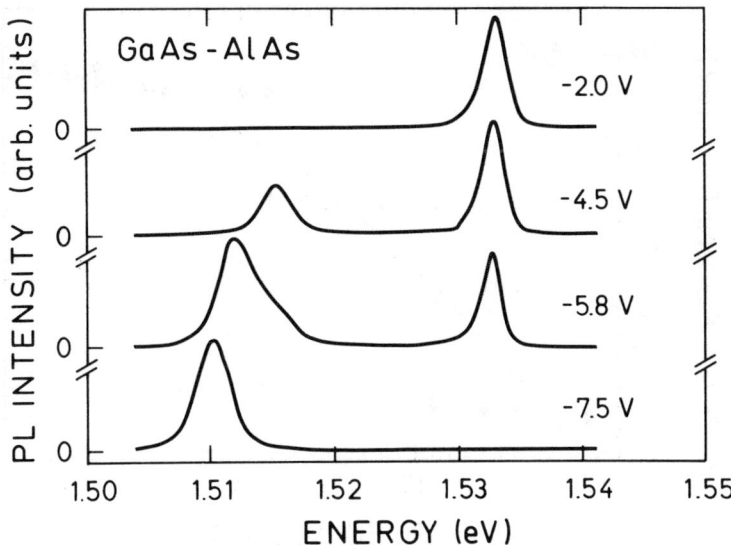

Fig. 5.17. Photoluminescence spectra of the undoped GaAs-AlAs superlattice with 14.4 nm GaAs wells, 3.4 nm AlAs barriers, and 50 periods at 5 K for different app-lied voltages in region A of Fig. 5.12. The excitation conditions are the same as in Fig. 5.12.

lines corresponding to F_3 and F_4, which cannot be assigned to a resonance field strength. The earlier assignment[16] as being due to the resonant coup-ling between C2 and C4 does not seem to make sense, since the second subband is not significantly occupied. Another possibility is the coupling between C1 in the well and the ground state X1 in the barrier, which should occur in this field range. However, more experiments are necessary to cla-rify the origin of this PL line. The assignment of the PL lines given above is a result of the comparison of the energetic positions of these lines with the single PL line observed at very low excitation densities, which exhibits the quadratic Stark shift. The dashed line in Fig. 5.18 denotes the energetic position of this single PL line. The dashed line and the symbols intersect at -2, -7.5 and -14 V. These voltages correspond to an energetic spacings of $\Delta E_i = (V_{BI} - V_{AP})/N$ of 70, 180, and 310 meV. The subband spacing at resonance $E_{Ci} - E_{C1}$ for this sample are 61, 163, and 308 meV for $i = 2$,

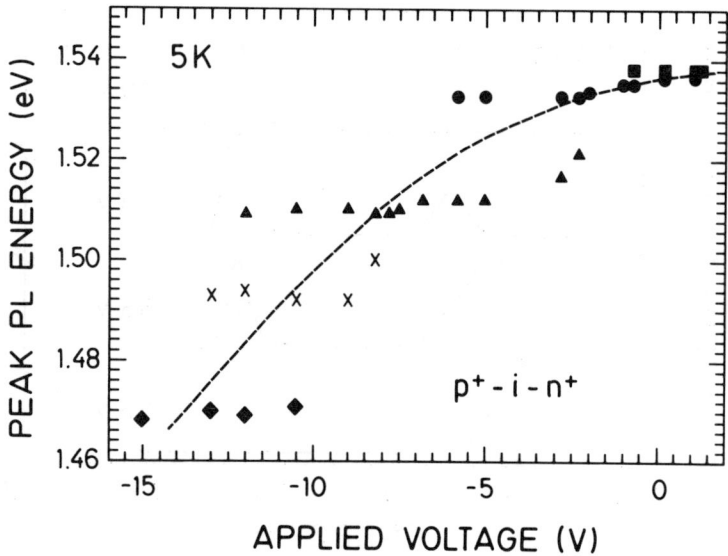

Fig. 5.18. Energetic positions of the PL lines in the undoped GaAs-AlAs superlattice with 14.4 nm GaAs wells, 3.4 nm AlAs barriers, and 50 periods at 5 K vs applied voltage. The excitation conditions are the same as in Fig. 5.12. The symbols denote different domain field strengths as discussed in the text. The dashed line denotes the Stark-shift of the PL line at low excitation conditions as measured by Tarucha et al.[51]

3, and 4, respectively. Therefore, the corresponding PL lines under domain formation are assigned to the field strengths in the vicinity of F_2, F_3, and F_4. A high-field domain becomes a low-field domain at larger electric fields. Therefore, the field strength increases from one plateau to the next from a value, which corresponds to non-resonant coupling, to a value, where the superlattice is resonantly coupled. This change of the field strength can be seen in Fig. 5.18 for F_2 and F_3, since the corresponding PL lines shift a little bit to lower energy, when F_2 and F_3 become a low-field domain.

In conclusion of this section, the presence of electric field domains has been directly shown using optical spectroscopy. Two regimes of well-defined field strengths exist. The low-field domain exhibits a field strength close to the resonance, while the high-field domain clearly experiences a field strength below resonance. This observation is in agreement with the simple picture based on current conservation and a drift velocity-field characteristic exhibiting several resonances. Electric field domains with resonant coupling up to C1→C4 have been observed.

5.4.2. *Intensity dependence of domain formation*

Domain formation in superlattices occurs only when a significant carrier density is present in the system to support the space charge layer at the domain boundary. Assuming that the domain boundary extends over one period, typical densities, which are necessary to support the field change at the domain boundary, are of the order of several 10^{11} cm^{-2}. Since a large fraction of photoexcited carriers disappears either via recombination or transport and therefore do not contribute to the space charge, a certain carrier density, i.e., power density, is required to produce stable domains. In order to demonstrate the existence of a critical power density, the PL signal in the undoped GaAs-AlAs superlattice with 14.4 nm wells, 3.4 nm barriers, and 50 periods has been measured as a function of laser intensity at an applied voltage, for which two PL lines are observed and the current

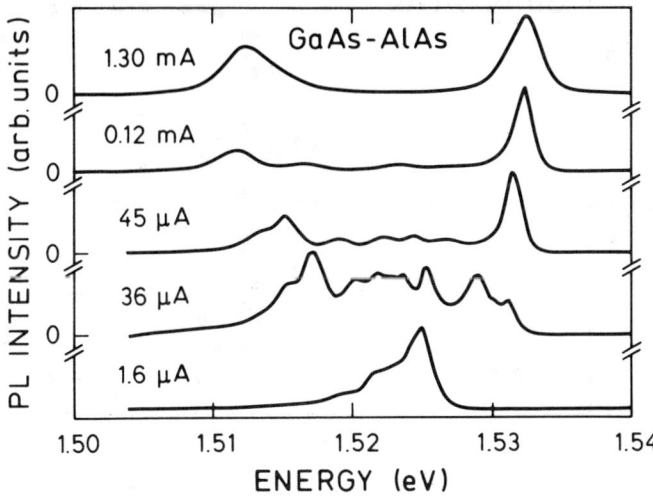

Fig. 5.19. Photoluminescence spectra of the undoped GaAs-AlAs superlattice with 14.4 nm GaAs wells, 3.4 nm AlAs barriers, and 50 periods at 5 K for different excitation intensities (1.916 eV or 647 nm) at an applied voltage of −5 V. The photocurrent at −8 V is used as a measure of the laser intensity (1 mA corresponds to about 50 W cm^{-2}).

exhibits jumps at large intensities. In Fig. 5.19 a series of selected PL spectra is shown at -5 V for different laser intensities. The photocurrent at -8 V was used as a measure of the light intensity. At very low intensities (1.6 μA) only a single PL line at an energy between the lines of domain F_2 (II) and F_3 (III) are observed. As the intensity is increased the PL line becomes broader with a lot of additional fine structure. At 45 μA photocurrent already two lines close to the energy of the domains II and III are seen with smaller structures in between. Further increase of the intensity establishes the PL lines of domain II and III even more, while the fine structure between the peaks disappears. At a photocurrent of 1.3 mA a similar PL spectrum as in Fig. 5.17 is observed and the fine structure has disappeared. The dependence of the PL spectrum on the laser intensity clearly reveals a threshold for domain formation.

In Fig. 5.20 the peak energies of the PL spectra measured as a function of the photocurrent at -8 V, which is a measure of the photoexcited carrier

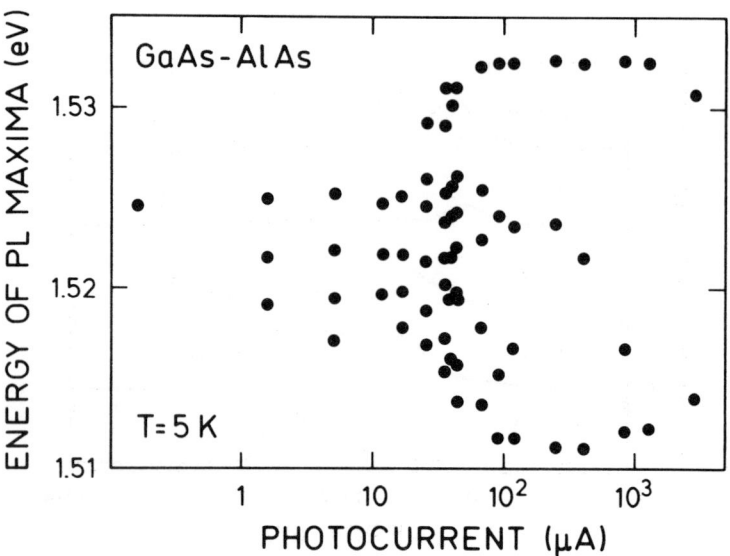

Fig. 5.20. Photoluminescence peak energy of the undoped GaAs-AlAs superlattice with 14.4 nm GaAs wells, 3.4 nm AlAs barriers, and 50 periods at -5 V vs the photocurrent measured at -8 V. The experimental conditions were identical to the ones of Fig. 5.19.

density, are plotted on semilogarithmic photocurrent scale. At very low and very high photocurrent only one and two peaks are observed, respectively. Between 10 and 100 μA the PL spectra display a very rich fine structure. The threshold for domain formation lies between 10 and 100 μA in this sample and is confined to a very small region of intensities. Stable domain formation is only achieved at very high intensities. In the threshold region the domains are not stable yet as will be discussed in the last section of this chapter.

5.4.3. *Spatial distribution*

Theoretically a spatial ordering of the domains is expected. The high-field domain will start from the contact at lower bias and then grow in size through the superlattice, until it reaches the contact at higher bias. In a p^+-i-n^+ structure under reverse bias the high-field domain should reside near the n-contact, while the low-field domain should be located near the p^+ contact. Photoluminescence spectroscopy at different laser energies, which changes the penetration depth of the excitation, can probe the spatial order of the domains. The PL intensity ratio of the two domains depends on the penetration depth of the light. The spatial positions of the two domains should be interchanged when one switches from reverse to forward bias. The spatial position of the domains leads to an identification of the space charge that resides at the domain boundary.

In Fig. 5.21(a) the ratio of the intensities of the PL from the high-field and low-field domain is plotted versus the reverse bias voltage for different excitation wavelengths. The inset shows a schematic diagram of the experimental configuration. At a constant voltage the high-field domain has a relatively lower PL intensity with respect to the low-field domain for excitation with 2.335 eV (531 nm) than for excitation with 1.916 eV (647 nm). This is true for the whole voltage region displayed. To observe the high-field domain with an excitation energy corresponding to a smaller penetration depth, a larger electric field is necessary implying that the high-field domain resides below the low-field domain in this bias configuration. As a result the high-field domain is located near the n-contact, while the low-field domain is closer to the p^+ contact layer. This leads to the conduction band edge distribution shown in the inset of Fig. 5.21(a). As a consequence the space-charge at the domain boundary are electrons. This is expected since the heavy holes, due to their larger effective mass, are much more localized

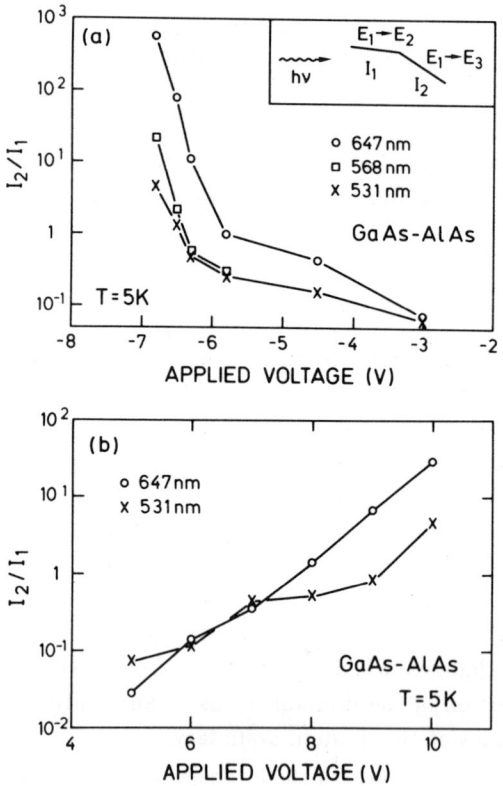

Fig. 5.21. Intensity ratio of PL lines of different domains as indicated in the inset vs applied voltage ((a) region A and (b) region D in Fig. 5.12) for the undoped superlattice with 14.4 nm GaAs wells, 3.4 nm AlAs barriers, and 50 periods at 5 K for different laser wavelengths. The laser intensity was adjusted in such a way that the photocurrent at −8 V was 1 mA, corresponding to a laser intensity of of about 50 W cm^{-2}.

than the electrons in this system and are confined to region near the p$^+$ contact. Therefore, in these PL experiments the holes are used as a detector for the electrons. The spatial ordering of the domains is expected to reverse when the applied voltage is changed from reverse to forward bias. In this case the high-field domain should be located near the p$^+$ contact and the low-field domain near the n-contact. In Fig. 5.21(b) the PL intensity ratio of the two existing domains in the forward bias region D of Fig. 5.12 is shown for two different excitation energies. The wavelength dependence is more complicated in this case and the identification of the ordering cannot be made unambiguously. However, if the relative intensities at 1.916 eV of reverse and forward bias are compared, the anticipated ordering can be proven. In Fig. 5.22 the PL intensity ratios of forward and reverse bias are shown in one graph. The voltage scale is relative to the respective resonance voltage for tunneling from C1 into C2, which occurs at -2 V for reverse bias and at 5.1 V for forward bias. For a fixed relative field the intensity ratio is about a factor of 2 to 3 larger in forward bias than in reverse. This implies that in forward bias the high-field domain must be closer to the front contact, i.e., the p$^+$ contact, than the low-field domain. The crossing of the two curves at high voltages is an artifact arising from the definition of the intensity ratio, since the intensity I_1 vanishes for resonant coupling C1$\rightarrow$C3. Assuming two domains with one boundary we conclude that the

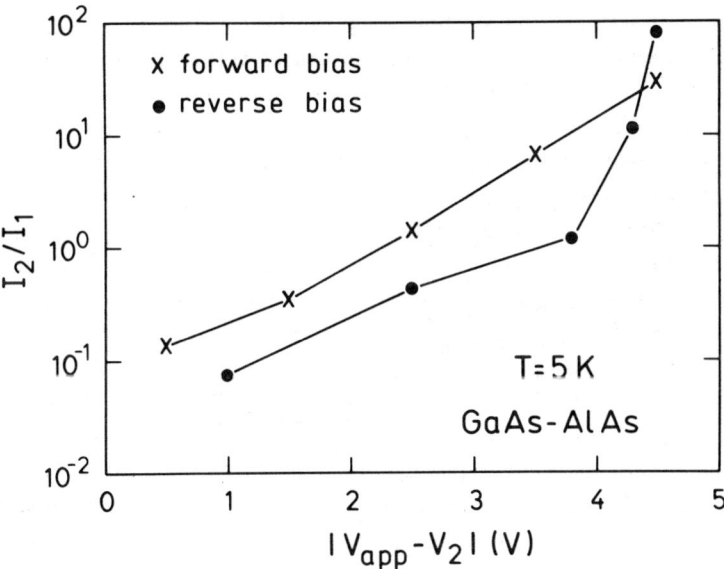

Fig. 5.22. Comparison of the PL intensity ratios originating from different domains in regions A and D of Fig. 5.12 in the undoped superlattice with 14.4 nm GaAs wells, 3.4 nm AlAs barriers, and 50 periods with an excitation energy of 1.916 eV (647 nm). The voltage scale is relative to $V_2 = \pm N(E_{C2} - E_{C1})/e + V_{BI}$, the plus and minus signs referring to forward and reverse bias.

high-field domain resides near the p^+ contact in forward bias. The ordering is therefore as expected within the simple model described in section 5.2. The calculation of the field distribution within the microscopic model supports this result.

Recently, the spatial arrangement of electric field domains in doped superlattices was investigated by cathodoluminescence (CL) in a scanning electron microscope (SEM).[55] The CL-spectra of the doped sample with 3×10^{17} cm^{-3} Si donors were recorded at a fixed applied voltage for different acceleration voltages of the SEM. This technique leads to a direct variation of the penetration depth of the exciting electron beam. For a voltage in the first plateau (-3.2 V) of Fig. 5.4(b) the relative intensity of the high-energy CL line decreases with increasing acceleration voltage, while the relative intensity of the low-energy CL line increases. This experiment directly confirms the spatial ordering with the high-field domain residing at the anode. Furthermore, because of the continuous increase of the low-energy line, i.e., the high-field domain, with increasing electron beam energy, the existence of a single domain boundary can be concluded. The experiment was also performed in the second plateau of this sample at -8.2 V (cf. Fig. 5.16). The dependence of the CL spectra on the electron beam energy is shown in Fig. 5.23. The qualitative behavior is similar to the one in the first plateau. However, due to the larger splitting of the CL lines, a quantitative analysis of these spectra was possible. The inset of this figure shows the relative integrated intensity of peak III with respect to peak II as obtained by fitting two Gaussian lines centered at 793.4 and 797.8 nm to the experimental spectra. The widths of the two peaks were kept fixed at 2.81 and 3.55 nm, respectively. In addition to the continuous increase of the relative intensity, the experimental data can be fitted applying the model of the modified Gaussian approximation.[53] As shown in Ref. 49 the only fitting parameters used are the spatial extent of the low-field domain and the domain boundary. The best fit is shown in the inset of Fig. 5.23 as a solid line using 18 ± 1 periods in the low-field domain and 2 ± 1 periods for the domain boundary. From the applied voltage it is expected that 17 ± 2 periods are in the low-field domain in excellent agreement with the value obtained from the CL experiments. The result of the spatial extent of the domain boundary is in very good agreement with the microscopic model by Prengel et al.[35−38], which predicts about two to three periods.

In summary, the domains are spatially ordered. The arrangement is

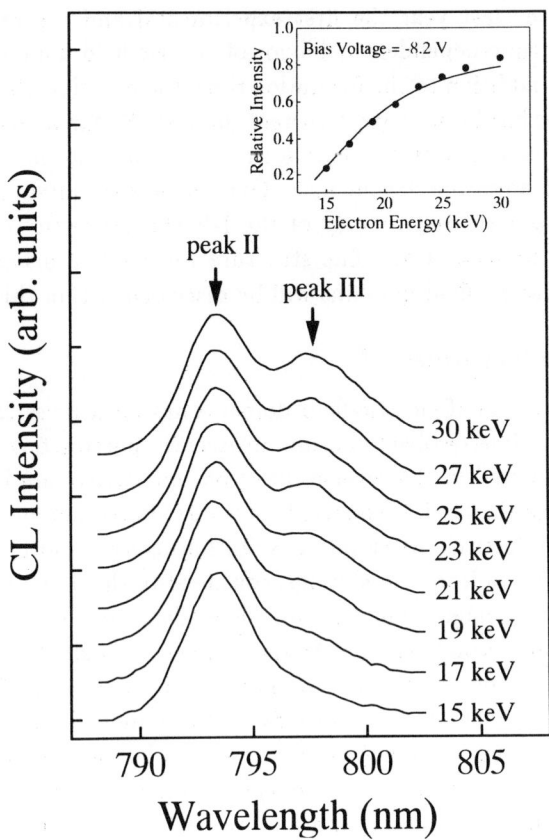

Fig. 5.23. Cathodoluminescence spectra in the doped superlattice with 9.0 nm GaAs wells, 4.0 nm AlAs barriers, 40 periods, and a Si-doping of 3×10^{17} cm^{-3} for an applied voltage of -8.2 V at 5 K using different electron beam energies. The CL lines associated with domains II and III are labelled by peak II and III, respectively. Inset: The relative intensity of peak III with respect to peak II vs electron beam energy (dots). The solid line is a fit to the data points (from Ref. 51).

always such that the low-field domain is located on the cathode side of the structure, while the high-field domain is situated on the anode side. Only one domain boundary exists separating the low- and high-field domains. This domain boundary is formed by electron accumulation.

5.5. Dynamics of domain formation

During the last year the first experimental and theoretical investigations of the time-dependent behavior of electric-field domains have been performed. In addition to the formation time, the question about the stability of domains has been of great current interest. So far we have been only discussing stable domains, i.e., the electric field domains and the boundary do not exhibit any time dependence. One unresolved question within this context has been the asymmetry of the I-V characteristic in Fig. 5.6(a), in particular the lack of any fine structure for positive biases. The time evolution of electric field domains will be discussed within this section.

5.5.1. *Formation time*

The formation of electric-field domains has been investigated in a p-i-n diode using time-resolved photoluminescence spectroscopy.[54] When the p-i-n diode is biased in the reverse direction, no carriers are injected from the contact, and the applied electric field is distributed homogeneously over the whole superlattice. After photoexcitation using a short, intense laser pulse, the evolution from the homogeneous field to the domain distribution can be studied by monitoring the time dependence of the PL spectrum in the pico- to nanosecond regime. At short times, a single, broad PL line is observed, which evolves into two PL lines as time progresses. This is shown in Fig. 5.24(a), where the PL spectra of the undoped superlattice with 14.4 nm GaAs wells, 3.4 nm AlAs barriers, and 50 periods are plotted for different time delays. In this voltage regime domains II and III coexist (cf. Fig. 5.18) and under domain formation two PL lines are expected. At short time (1 ns) only a broad line is observed, which develops with increasing time delay into two well-resolved PL lines. The PL signal of both lines decays exponentially within a few nanoseconds as shown in Fig. 5.24(b). The time constants are 1.5 ns for both lines. More important is, however, the time scale for the formation of the domains. After photoexcitation a certain time scale is required in order to accumulate the necessary charge at the domain boundary to establish the two field regions of the domains. From Fig.5.24 a formation time of 3 ns is deduced, although the transport of electrons and holes through the whole superlattice occurs on a much longer time scale. For domains III and IV the formation time is even shorter (1-2 ns[54]), since the transport time decreases with increasing applied electric field. Further investigations are necessary to clearly identify the time scale

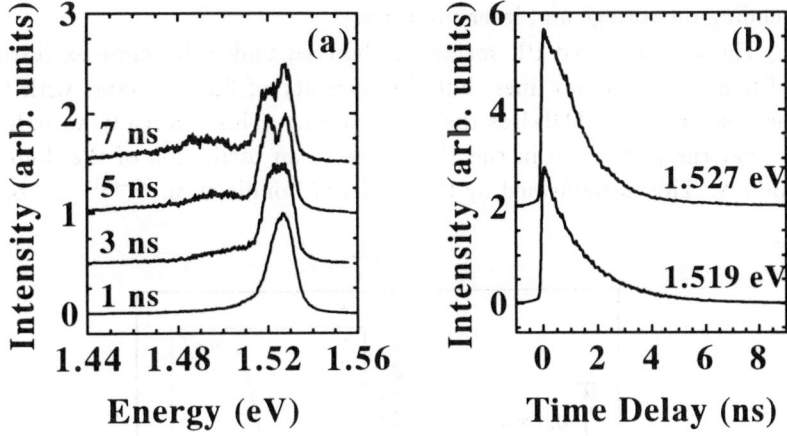

Fig. 5.24. (a) Photoluminescence spectra at different delay times in the undoped super-lattice with 14.4 nm GaAs wells, 3.4 nm AlAs barriers and 50 periods for an applied voltage of −3 V at 5.6 K. The spectra have been normalized and shifted. (b) Temporal evolution of the intensity of the two PL lines in (a). The upper curve has been shifted by 2 units.

for the formation. Recent theoretical investigations, however, also reveal a rather short time scale for the formation of the domains in comparison with the transport time through the whole superlattice.[38]

5.5.2. *Damped oscillations of the photocurrent*

The dynamical behavior of an undoped superlattice subject to step-like photoexcitation was recently investigated experimentally[40,55] and theoretically.[39] The experiments were performed on a p-i-n diode containing an undoped superlattice with 9.0 nm GaAs wells, 4.0 nm AlAs wells, and 40 periods. When the system is biased in the second plateau, i.e., domains II and III, the photocurrent (PC) exhibits damped oscillations for a step-like excitation. The period of the oscillations is typically in the 100 ns regime, but depends on the laser intensity and applied voltage. The dependence on the applied voltage is shown in Fig. 5.25 for an intensity of 250 Wcm^{-2}. At considerably larger and lower intensities, the PC response becomes also step-like, i.e., the damped oscillations appear only within a certain window of intensities. It is also apparent from Fig. 5.25 that the oscillation is not always purely harmonic, but sometimes contains higher harmonics depending on voltage and laser intensity.

The time-resolved PL spectrum observed under the same excitation conditions exhibits two lines, but the intensity of line oscillates with the same frequency (4-8 MHz) as the photocurrent. This is already an indication that the oscillation in the PC is due to an oscillation of the domain boundary. This is confirmed by the model of Bonilla et al.,[39-41] who have

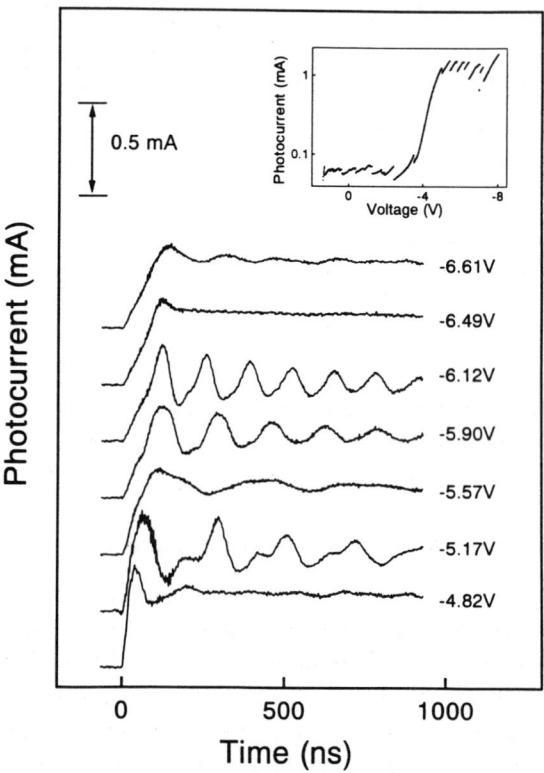

Fig. 5.25. Measured photocurrent vs time in the undoped superlattice with 9.0 nm GaAs wells, 4.0 nm AlAs barriers, and 40 periods at 18 K for different applied voltages using 250 Wcm^{-2} at 654 nm. The different traces have been offset vertically for clarity (from Ref. 55).

calculated the time dependence of the photocurrent using a discrete Poisson, continuity, and rate equation in one dimension with appropriate boundary conditions. The tunneling probabilities are taken into account by using the drift velocity-field characteristic of the sample at low excitation densities in the continuity equation. The result is shown in Fig. 5.26 for different excitation densities. Both the current density and the time scale have been renormalized and appear as dimensionless variables. The typical behavior of the experimentally observed intensity, however, is reproduced. Furthermore, the origin of the oscillations can be clearly identified. The domain wall is created on a relatively short time scale. Due to the existence of multiple steady states, however, the domain boundary oscillates back and forth for a few periods about a fixed value of the position. The electric fields within the two domains also oscillate for a few periods as expected from the discussion in the section on multistability.

This oscillatory behavior is rather interesting, since it has important

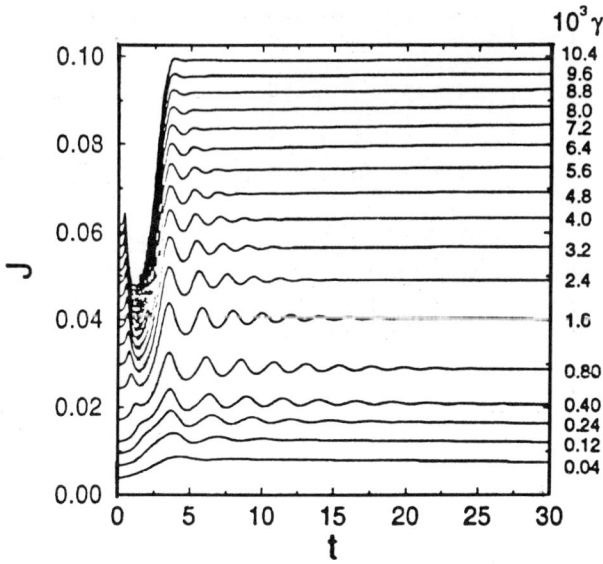

Fig. 5.26. Evolution of the photocurrent J for different values of the photogeneration γ increasing from the bottom to the top keeping the voltage constant (from Ref. 39).

consequences on the phase diagram, in which the photoexcitation density is plotted versus the applied voltage (cf. Fig. 5.27). An important conclusion drawn from this investigation is the existence of a *unstable* regime in this phase diagram, i.e., there is a wide range of values of the carrier density and applied voltage, where the system exhibits oscillatory behavior. Although we have only be discussing damped oscillations in photoexcited superlattices so far, it is obvious that these conclusions can probably also applied to doped superlattices, if the carrier density is reduced. However, the observation of this regime in doped superlattice is rather difficult, since

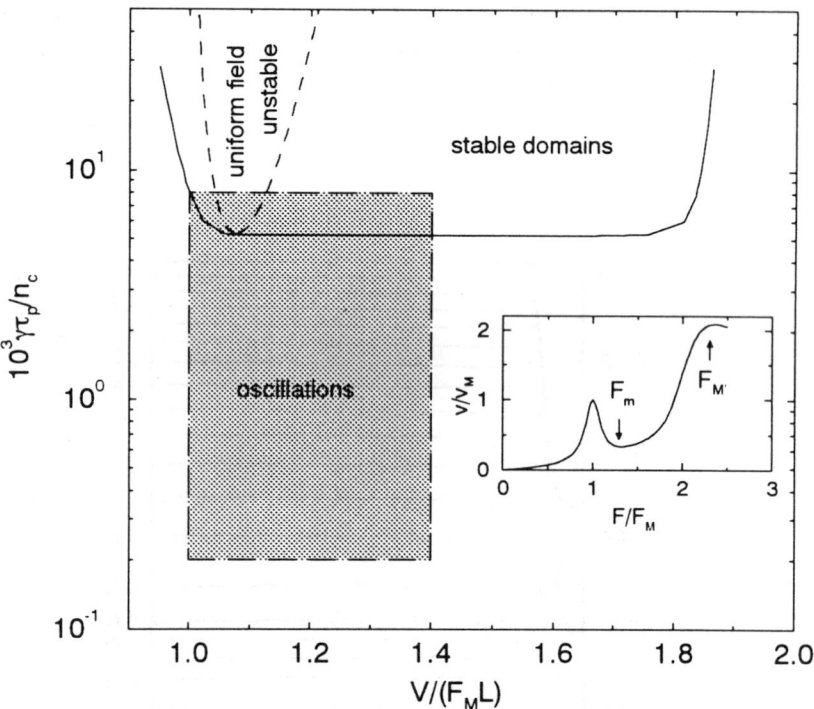

Fig. 5.27. Theoretical phase diagram of the photogeneration γ vs applied voltage V in dimensionless units. The shaded region denotes the approximate range, where the domain boundary oscillates. The continuous (dotted) curve delimits the region for which the domain (uniform-field) solution is stable (unstable). Inset: Normalized field dependence of the carrier velocity used as input in the calculations (from Ref. 39).

a number of samples with different doping levels has to be grown in order to investigate the oscillatory regime.

The observation of damped Gunn oscillations induced by intense pico-second light pulses in strongly coupled superlattices[56] should be reexamined in view of the observation of self-sustained current oscillations and damped oscillations in photoexcited superlattices.

5.5.3. *Self-oscillations of the current*

In this section we will briefly discuss some very recent results on self-oscillations in doped superlattices.[57] The I-V characteristics of the super-

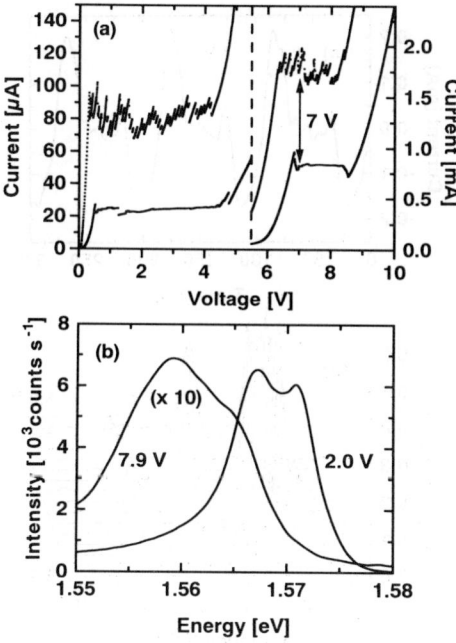

Fig. 5.28. (a) I-V characteristics of the doped superlattice with 9.0 nm GaAs wells, 4.0 nm AlAs barriers, 40 periods, and a Si-doping of 3×10^{17} cm^{-3} at 5 K. The upper curve has been recorded with a laser power of 500 mW at 1.653 eV. The left half of the plot has been multiplied by a factor of 16. (b) PL spectra at 5 K for two applied voltages in the two plateau regions at very low excitation power of 10 μW. The spectrum at 7.9 V has been rescaled by a factor of 10 (from Ref. 57).

lattice with 9.0 nm GaAs wells, 4.0 nm AlAs wells, 40 periods, and a Si-doping of 3×10^{17} cm^{-3} shows a very peculiar asymmetry (cf. Fig. 5.6(a)). For a positive applied voltage no fine structure is observed, unless the sample is strongly photoexcited. This is shown in Fig. 5.28(a), where the I-V characteristic is plotted for positive applied voltages in the dark and under strong photoexcitation. In Fig. 5.28(b) two representative PL spectra from each plateau region are shown clearly indicating the existence of two well-defined electric field regions, i.e., domains. However, the current does not exhibit the typical fine structure as usually observed under domain formation. The puzzle is solved, when the current is recorded on a much shorter time scale. As shown in Fig. 5.29 (a) for an applied voltage of 7 V, the

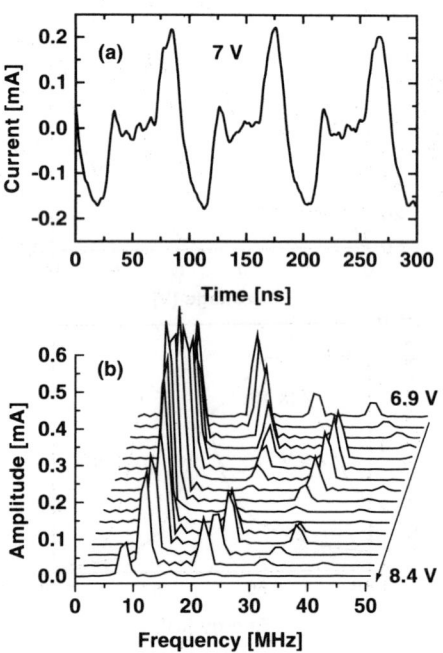

Fig. 5.29. (a) Typical trace of the time-resolved current at 7 V with a DC average current of 0.8 mA in the doped superlattice with 9.0 nm GaAs wells, 4.0 nm AlAs barriers, 40 periods, and a Si-doping of 3×10^{17} cm^{-3} at 5 K. (b) Fourier transform of the dark current oscillations vs frequency for different applied voltages (0.1 V steps) in the upper plateau (from Ref. 57).

current exhibits self-oscillations in this bias direction with a frequency in the 10 MHz regime, which lead in the time-averaged I-V characteristic to a completely flat current. In the first plateau the system also oscillates, but with a lower frequency of 250 kHz. In Fig. 5.29(b) a series of Fourier transform of the current oscillations are plotted for applied voltages in the upper plateau regime without illuminating the sample. The Fourier transforms clearly demonstrate that the current oscillations contain in addition to the fundamental frequency of 11 MHz several higher harmonics. While the positions of the frequency peaks do not vary much with bias, the intensity distribution among the peaks changes rather drastically. The observation of higher harmonics in the Fourier spectra is not surprising, but rather reflects the non-linearity of the system. However, it is interesting to note that the oscillations are close to being harmonic near the center of the plateau region (7.9 V).

5.6. Γ-X tunneling and domain formation

Recently, the appearance of domain formation due to resonant tunneling from the Γ-minimum in the well into the X-minimum in the barrier material was discussed.[58] The authors used a sample with 7.5 nm GaAs well and 4.0 nm AlAs barriers. In this system the second subband is already above the X-minimum in the barrier. We also believe that some of the unresolved questions above can be related to the resonant transfer from the Γ-minimum in GaAs to the X-minimum in AlAs. The high-field domain in the second plateau in the superlattice with 9.0 nm GaAs wells and the appearance of three PL peaks in the fourth plateau in the sample with 14.4 nm GaAs wells are also tentatively connected to the Γ-X transfer. However, more research is necessary to unambiguously identify the true origin of these features.

5.7. Summary and conclusion

Electric-field domain formation in semiconductor superlattices is particularly important with regard to any device applications, since it usually requires a certain amount of doping. The dynamical behavior of electric-field domains demonstrates the widely spread existence of oscillatory behavior. Stable domains are only formed above a critical carrier density. However, the oscillations can persist to rather low carrier densities depending on the coupling between adjacent wells. The multistability of the I-V characteristics could lead to some device applications, if stable electric-field domains can be produced at room temperature, which is currently not the case. The field of stable domains has matured in the last few years. However, we are only beginning to realize the potential and unresolved questions with regard to the dynamical behavior.

5.8. Acknowledgments

It is a great pleasure to thank all collaborators in this field over the last six years, L.L. Bonilla, J. Kastrup, R. Klann, K. von Klitzing, S.H. Kwok, B. Laikhtman, R. Merlin, W. Müller, K. Ploog, F. Prengel, H. Schneider, E. Schöll, A. Wacker, and, in particular, A. Fischer, who grew a number exceptional samples.

References

1. B.K. Ridley and T.B. Watkins, *Proc. Phys. Soc.* **78**, 293 (1961).
2. J.B. Gunn, *Solid State Commun.* **1**, 88 (1963).
3. S.M. Sze, *Physics of Semiconductor Devices*, 2nd edition (Wiley, New York, 1981), p.641.
4. K.W. Böer and P. Voss, *Phys. Rev.* **171**, 899 (1968).
5. L. Esaki and L.L. Chang, *Phys. Rev. Lett.* **33**, 495 (1974).
6. Y. Kawamura, K. Wakita, H. Asaki, and K. Kurumada, *Jpn. J. Appl. Phys.* **25**, L928 (1986).
7. Y. Kawamura, K. Wakita, and K. Oe, *Jpn. J. Appl. Phys.*, **26**, L1603 (1987).
8. T. Furuta, K. Hirakawa, J. Yoshino, and H. Sakaki, *Jpn. J. Appl. Phys.* **25**, L151 (1986).
9. K.K. Choi, B.F. Levine, R.J. Malik, J. Walker, and C.G. Bethea, *Phys. Rev. B* **35**, 4172 (1987).
10. K.K. Choi, B.F. Levine, C.G. Bethea, J. Walker, and R.J. Malik, *Appl. Phys. Lett.* **50**, 1814 (1987).
11. H.S. Newman and S.W. Kirchoefer, *J. Appl. Phys.* **62**, 706 (1987).
12. K.K. Choi, B.F. Levine, N. Jarosik, J. Walker, and R. Malik, *Phys. Rev. B* **38**, 12362 (1988).
13. E.S. Snow, S.W. Kirchoefer, and O.J. Glembocki, *Appl. Phys. Lett* **54**, 2023 (1989).
14. H.T. Grahn, H. Schneider, and K. v. Klitzing, *Appl. Phys. Lett.* **54**, 1757 (1989).
15. M. Helm, P. England, E. Colas, F. DeRosa, and S.J. Allen, Jr., *Phys. Rev. Lett.* **63**, 74 (1989).

16. H.T. Grahn, H. Schneider, and K. v. Klitzing, *Phys. Rev. B* **41**, 2890 (1990).
17. M. Helm, J.E. Golub, and E. Colas, *Appl. Phys. Lett.* **56**, 1356 (1990).
18. H.T. Grahn, R.J. Haug, W. Müller, and K. Ploog, *Phys. Rev. Lett.* **67**, 1618 (1991).
19. S.H. Kwok, E. Liarokapis, R. Merlin, and K. Ploog, in *Light Scattering in Semiconductor Structures and Superlattices*, edited by D. J. Lockwood and J. F. Young (Plenum, New York, 1991), p. 491.
20. H.T. Grahn, W. Müller, K. v. Klitzing and K. Ploog, *Surf. Sci.* **267**, 579 (1992).
21. I. Gravé, A. Shakouri, N. Kuze, and A. Yariv, *Appl. Phys. Lett.* **60**, 2362 (1992).
22. A. Shakouri, I. Gravé, Y. Xu, A. Ghaffari, and A. Yariv, *Appl. Phys. Lett.* **63**, 1101 (1993).
23. S. Murugkar, S.H. Kwok, G. Ambrazevičius, H.T. Grahn, K. Ploog, and R. Merlin, *Phys. Rev. B* **49**, 16849 (1994).
24. S.H. Kwok, R. Merlin, H.T. Grahn, and K. Ploog, *Phys. Rev. B* **50**, 2007 (1994).
25. J. Kastrup, H.T. Grahn, K. Ploog, F. Prengel, A. Wacker, and E. Schöll, *Appl. Phys. Lett.* **65**, 1808 (1994).
26. R.E. Cavicchi, D.V. Lang, D. Gershoni, A.M. Sergent, H. Temkin, and M.B. Panish, *Phys. Rev. B* **38**, 13474 (1988).
27. T.H.H. Vuong, D.C. Tsui, and W.T. Tsang, *Appl. Phys. Lett.* **52**, 981 (1988).
28. T.H.H. Vuong, D.C. Tsui, and W.T. Tsang, *J. Appl. Phys.* **66**, 3688 (1989).
29. R.A. Suris, *Sov. Phys. Semicond.* **7**, 1030 (1974).
30. R.A. Suris, *Sov. Phys. Semicond.* **7**, 1035 (1974).
31. A.A. Ignatov, V.I. Piskarev, and V.I. Shashkin, *Sov. Phys. Semicond.* **19**, 1345 (1985).
32. B. Laikhtman, *Phys. Rev. B* **44**, 11260 (1991).
33. B. Laikhtman and D. Miller, *Phys. Rev. B* **48**, 5395 (1993).
34. D. Miller and B. Laikhtman *Phys. Rev. B* **50**, in press (1994).
35. F. Prengel, A. Wacker, and E. Schöll, *Phys. Rev. B* **50**, 1705 (1994).
36. A. Wacker, F. Prengel, and E.Schöll, in *Proceedings of the 22nd International Conference on the Physics of Semiconductors*, edited by D.J. Lockwood (World Scientific, Singapore, 1994), in press.
37. E. Schöll and A. Wacker, in *Nonlinear Dynamics and Pattern Formation in Semiconductors and Devices*, edited by F.J. Niedernostheide (Springer Verlag, Berlin, 1994), in press.
38. F. Prengel, Diplomarbeit, Technische Universität Berlin (1994).
39. L.L. Bonilla, J. Galán J.A. Cuesta, F.C. Martínez, and J.M. Molera, *Phys. Rev. B* **50**, 8644 (1994).

40. R. Merlin, S.H. Kwok, T.B. Norris, H.T. Grahn, K. Ploog, L.L. Bonilla, J. Galán, J.A. Cuesta, F.C. Martínez, and J.M. Molera, in *Proceedings of the 22nd International Conference on the Physics of Semiconductors*, edited by D.J. Lockwood (World Scientific, Singapore, 1994), in press.

41. L.L. Bonilla, in *Nonlinear Dynamics and Pattern Formation in Semiconductors and Devices*, edited by F.J. Niedernostheide (Springer Verlag, Berlin, 1994), in press.

42. A.N. Korotkov, D.V. Averin, and K.K. Likarev, *Appl. Phys. Lett.* **62**, 3282 (1993).

43. Y. Zhang, Y. Li, D. Jiang, X. Yang, and P. Zhang, *Appl. Phys. Lett.* **64**, 3416 (1994).

44. C.B. Duke, in *Tunneling Phenomena in Solids*, edited by E. Burstein and S. Lundqvist (Plenum Press, New York, 1969), p.31.

45. D.A.B. Miller, D.S. Chemla, T.C. Damen, A.C. Gossard, W. Wiegmann, T.H. Wood, and C.A. Burrus, *Phys. Rev. Lett.* **53**, 2173 (1984).

46. H.T. Grahn, H. Schneider, W.W. Rühle, K. v. Klitzing, and K. Ploog, *Phys. Rev. Lett.* **64**, 2426 (1990).

47. D. Bertram, H. Lage, H.T. Grahn, and K. Ploog, *Appl. Phys. Lett.* **64**, 1012 (1994).

48. R. Klann, H.T. Grahn, and K. Ploog, *Phys. Rev. B* **50**, 15. Oct. (1994).

49. S.H. Kwok, H.T. Grahn, K. Ploog, and R. Merlin, *Phys. Rev. Lett.* **69**, 973 (1992).

50. S.H. Kwok, H.T. Grahn, M. Ramsteiner, K. Ploog, F. Prengel, A. Wacker, E. Schöll, S. Murugkar, and R. Merlin, *submitted to Phys. Rev. B* (1994).

51. S. Tarucha and K. Ploog, *Phys. Rev. B* **38**, 4198 (1988).

52. S.H. Kwok, U. Jahn, J. Menniger, H.T. Grahn, and K. Ploog, *submitted to Appl. Phys. Lett.* (1994).

53. C.J. Wu and D.B. Wittry, *J. Appl. Phys.* **49**, 2827 (1978).

54. H.T. Grahn, R. Klann, and W.W. Rühle, in *Proceedings of the 22nd International Conference on the Physics of Semiconductors*, edited by D.J. Lockwood (World Scientific, Singapore, 1994), in press.

55. S.H. Kwok, T.B. Norris, L.L. Bonilla, J. Galán, J.A. Cuesta, F.C. Martínez, J.M. Molera, H.T. Grahn, K. Ploog, and R. Merlin, *submitted to Phys. Rev. Lett.* (1994).

56. H. Le Person, C. Minot, L. Boni, J.F. Palmier, and F. Mollot, *Appl. Phys. Lett.* **60**, 2397 (1992).

57. J. Kastrup, R. Klann, H.T. Grahn, K. Ploog, R. Merlin, L.L. Bonilla, J. Galán, J.A. Cuesta, F.C. Martínez, and J.M. Molera, *in preparation* (1994).

58. Y. Zhang, Y. Yang, W. Liu, P. Zhang, and D. Jiang, *Appl. Phys. Lett.* **65**, 1148 (1994).